Fundamentals of Plant Pathology

NIPA® GENX ELECTRONIC RESOURCES & SOLUTIONS P. LTD.
New Delhi-110 034

About the Authors

Dr. Vikram Mahipatrao Gholve, a distinguished Professor in the Department of Plant Pathology at Vasantrao Naik Marathwada Krishi Vidyapeeth (VNMKV), Parbhani, is a renowned expert in biological disease management. His research focuses on microbial consortia and the management of fungal, bacterial, and viral diseases affecting various crops. With an impressive academic portfolio that includes 57 research papers, two authored books, one edited book, numerous academic manuals, and several prestigious awards, Dr. Gholve has significantly advanced the field of plant pathology. Notably, he played a pivotal role in the ICRISAT-funded Harvest-Plus Project, which aimed to enhance sorghum production, and contributed to the development and release of five prominent sorghum varieties with state, zonal, and national relevance.

Pawar Ganesh Shahaji is a Ph.D. scholar in Plant Pathology at Vasantrao Naik Marathwada Krishi Vidyapeeth (VNMKV), Parbhani, with a Master's degree in Agriculture. He has served as a Subject Matter Specialist at ICAR-KVK, Assistant Professor at Dr. BSKKV, Dapoli, and as a Research Assistant in the Post Entry Quarantine project. Pawar has authored 25 research papers, 4 book chapters, and one edited book. His research focuses on crop diseases, pest management, and bio-agent efficacy, earning him three Best Presentation Awards at international conferences. He actively participates in field trials, diagnostic visits, and farmer training programs as guest speaker and radio talks.

Fundamentals of Plant Pathology

Concepts, Causes, and Control Strategies

Vikram Mahipatrao Gholve
Professor (CAS)
Department of Plant Pathology
Vasantrao Naik Marathwada Krishi Vidyapeeth (VNMKV)
Parbhani-431402, Maharashtra, India

Ganesh Shahaji Pawar
Department of Plant Pathology
Vasantrao Naik Marathwada Krishi Vidyapeeth (VNMKV)
Parbhani-431402, Maharashtra, India

NIPA® GENX ELECTRONIC RESOURCES & SOLUTIONS P. LTD.
New Delhi-110 034

NIPA® GENX ELECTRONIC RESOURCES & SOLUTIONS P. LTD.

101,103, Vikas Surya Plaza, CU Block
L.S.C. Market, Pitam Pura, New Delhi-110 034
Ph : +91-11-43860225, Mob.: +91 9717133558, 9540816132
E-mail: newindiapublishingagency@gmail.com
Website: www.nipaersources.com

Print ISBN: 978-93-58875-59-1
ebook ISBN: 978-93-58873-84-9

Composed and Designed by NIPA®.

Foreword

It gives me great pleasure to write the foreword for the book ***Fundamentals of Plant Pathology***, authored by **Dr. Vikram Gholve** and **Mr. Ganesh Pawar**. This book is a timely and valuable contribution to agricultural education, offering a foundational understanding of plant pathology for undergraduate students.

Plant pathology is a vital discipline that underpins sustainable crop production by addressing plant health, disease diagnosis, and effective management strategies. In an era where climate change, intensive agriculture, and emerging pathogens pose increasing threats to global food security, a solid grounding in plant pathology is essential for future agronomists, researchers, and practitioners. This textbook addresses these needs comprehensively, with clarity and precision.

The authors have presented the subject matter in a lucid and structured manner, covering historical developments, types of plant pathogens, disease cycles, host-pathogen interactions, and integrated disease management approaches. The inclusion of recent advances and practical aspects reflects a well-balanced effort to bridge theoretical knowledge with field applicability.

I congratulate the authors for their dedicated and scholarly work and believe this textbook will serve as a valuable resource in agricultural colleges across the country.

I am confident that *Fundamentals of Plant Pathology* will enrich the academic journey of its readers and inspire them to contribute meaningfully to the field of plant health management.

Dr. B.V. Asewar
Director of Instruction & Dean,
F/A,VNMKV,Parbhani

Preface

The restructuring of agricultural education under the **6th Dean's Committee**, in alignment with the **National Education Policy (NEP) 2020**, aims to enhance the curriculum with a multidisciplinary and practical approach. The course **Fundamentals of Plant Pathology** has been developed to provide students with a foundational understanding of plant diseases, their causes, and management. Plant pathology plays a crucial role in ensuring crop health and productivity by studying the interactions between plants and various pathogens, including fungi, bacteria, viruses, and other microorganisms. This course introduces students to the concepts of plant disease classification, disease development, and different management strategies.

The course is structured to offer both theoretical and practical insights into plant pathology. The theoretical component covers key topics such as the history of plant pathology in India, disease classification, pathogen morphology and reproduction, and the principles of disease management. Students will gain knowledge about different plant pathogens and their role in plant health, along with methods to diagnose and control plant diseases. The practical sessions provide hands-on experience with microscopy, pathogen isolation, staining techniques, and the preparation of culture media. Additionally, students will be trained in plant disease diagnosis using Koch's postulates and the formulation and application of fungicides.

This book is designed as a comprehensive guide for undergraduate students, researchers, and academicians in the field of plant pathology. It integrates theoretical knowledge with practical applications to build competence in disease identification and management. The inclusion of suggested readings provides students with access to authoritative sources for further study. We sincerely acknowledge the efforts of experts and academicians who have contributed to shaping this course. We hope this book serves as a valuable resource, enabling students to develop the skills necessary for addressing plant disease challenges and contributing to sustainable agricultural practices.

Authors

Acknowledgement

We express our heartfelt gratitude to all those whose guidance and support have contributed to the successful completion of this book, Fundamentals of Plant Pathology. This work is the outcome of a collective effort inspired by the National Education Policy (NEP) 2020 and the academic restructuring recommended by the 6th Dean's Committee, aimed at enriching agricultural education through a multidisciplinary and application-oriented approach.

We are profoundly thankful to **Prof. Indra Mani**, Hon'ble Vice-Chancellor, Vasantrao Naik Marathwada Krishi Vidyapeeth, Parbhani (Maharashtra), for his constant encouragement, visionary leadership, and unwavering support throughout the development of this book. His commitment to academic excellence has been a guiding force in this endeavour.

We are also deeply indebted to the **Dr. B.V. Asewar**, Director of Instruction & Dean (F/A), VNMKV, Parbhani, for his insightful guidance and inspiring Foreword, which sets the academic tone of this work. His support and motivation have been instrumental in aligning this book with contemporary educational reforms.

Our sincere thanks are extended to **Dr. Syed Ismail**, Associate Dean & Principal College of Agriculture, Parbhani, for his valuable support, constructive feedback, and facilitation throughout the preparation of this manuscript.

We gratefully acknowledge the contributions of numerous **academicians, subject matter experts, and colleagues** whose expertise helped refine the content and ensure its academic rigor. Special appreciation is due to the educators and researchers who reviewed the manuscript and shared thoughtful suggestions that enhanced the overall quality of the book. We also thank the **students and research scholars** whose curiosity and eagerness to learn inspired the practical components of this book. Their enthusiasm underscored the need for a resource that bridges theory and application in plant pathology.

Finally, we are indebted to our families for their patience, understanding, and steadfast support, which made it possible to devote the necessary time and effort to this academic pursuit.

We hope this book serves as a valuable resource for students, researchers, and academicians, and contributes meaningfully to the field of plant pathology and sustainable agriculture

Authors

Contents

1

Introduction to Plant Pathology

Abstract

Plant pathology, or the study of plant diseases, is vital to understanding the impact of pathogens and environmental conditions on plant health. This chapter delves into the definition, scope, and historical development of plant pathology, highlighting its significance in agricultural productivity and food security. With roots tracing back to early human societies, plant pathology became a formal scientific discipline in the 19th century with the identification of microorganisms as the cause of plant diseases. The Bengal Famine of 1943 and the Coffee Rust Epidemic of the 1860s are discussed as key events that shaped the field. The chapter also addresses the economic importance of plant disease management, detailing its effects on crop yields and trade. Furthermore, it emphasizes the role of plant pathology in sustainable agriculture, advocating for disease-resistant crops and integrated management practices in response to global challenges like climate change and population growth. Ultimately, this chapter underscores plant pathology's crucial contribution to ensuring the health of crops and the security of food supplies in an ever-changing world.

Keywords: *Plant Pathology, Pathogens, Plant Diseases, Integrated Disease Management*

Objectives of Plant Pathology

The key objectives of plant pathology include:

1. Understanding Disease Etiology: Identifying the causal organisms or factors responsible for plant diseases.
2. Disease Diagnosis: Developing tools and techniques for early diagnosis of plant diseases.
3. Pathogenesis: Studying how pathogens cause disease in plants and how plants respond to infection.
4. Disease Management: Creating effective and sustainable strategies to control and prevent plant diseases.

Definition and Scope of Plant Pathology

Plant pathology, also known as phytopathology, is the study of plant diseases caused by pathogens (infectious agents) and environmental conditions (non-infectious agents). It encompasses the diagnosis of plant diseases, the mechanisms of disease development, and the development of management strategies. A widely accepted definition of plant pathology is provided by George N. Agrios (2005), who defined it as:

"Plant pathology is the study of the living entities and environmental conditions that cause diseases in plants, the mechanisms by which they induce disease in plants, the interactions between the plant and the pathogen, and the methods of preventing or managing plant diseases."

Plant pathology integrates principles from microbiology, botany, mycology, bacteriology, virology, and genetics to address plant health issues. It is a critical field for ensuring food security, agricultural sustainability, and the stability of ecosystems.

A Historical Perspective on Plant Pathology

The study of plant diseases is not new it has deep historical roots. Early human societies observed and recorded plant health issues, but it wasn't until the 19th century that plant pathology emerged as a formal scientific discipline. This shift occurred when researchers identified microorganisms as the underlying cause of many plant diseases, marking a turning point in how these problems were understood and addressed.

Several key events in history have profoundly influenced the development of plant pathology. Two of the most notable are the Bengal Famine of 1943 and the 19th-century coffee rust outbreak in Sri Lanka.

The Bengal Famine (1943)

One of the darkest chapters in India's agricultural history, the Bengal famine was fueled, in part, by crop failures linked to plant disease. Rice crops were devastated by brown spot, a fungal infection caused by *Cochliobolus miyabeanus* (formerly known as *Helminthosporium oryzae*). This contributed significantly to the severe food shortages that led to widespread starvation. The tragedy underscored the urgent need for scientific plant disease management to avert similar humanitarian crises.

The Coffee Rust Epidemic in Sri Lanka

In the 1860s, Sri Lanka then Ceylon was rocked by a catastrophic outbreak of coffee rust, caused by the fungus *Hemileia vastatrix*. The disease spread rapidly across plantations, crippling the once-thriving coffee industry. In response, many growers shifted to cultivating tea, which eventually replaced coffee as the island's dominant export crop. This epidemic highlighted the dangers of relying too heavily on a single crop species and emphasized the importance of disease-resistant farming systems.

Scope of Plant Pathology

1. **Disease Diagnosis:** Early and precise detection of plant pathogens is essential to prevent small infections from becoming widespread epidemics. Techniques such as polymerase chain reaction (PCR), enzyme-linked immunosorbent assay (ELISA), and high-throughput sequencing enable identification of pathogens including viruses, bacteria, and fungi before visible symptoms appear. Rapid diagnostics reduce unnecessary pesticide applications and allow targeted treatments, cutting costs and environmental harm.
2. **Etiology and Epidemiology:** Plant pathogens ranging from fungal rusts to bacterial blights interrupt physiological processes, resulting in yield and quality losses. Globally, between 20 percent and 40 percent of crop production is lost annually to pests and diseases, representing more than USD 220 billion in economic impact. In India, staple crops such as rice and wheat suffer 10–30 percent losses due to disease, highlighting the regional importance of etiological research.
3. **Integrated Disease Management:** Integrated Disease Management (IDM) weaves together biological, chemical, and cultural strategies to maintain disease below damaging levels. Biological controls—such as antagonistic microbes work alongside minimal, targeted fungicide use, crop rotation, and sanitation to suppress pathogens sustainably. IDM reduces reliance on broad-spectrum chemicals, preserving beneficial organisms and lowering resistance risks.
4. **Breeding for Resistance:** Genetic resistance offers a cost-effective, environmentally sound defense against pathogens. Resistance genes (R-genes) in crops trigger defense responses upon pathogen attack, reducing disease incidence. The success of Bt cotton and disease-resistant wheat varieties in India exemplifies how breeding efforts can stabilize yields and reduce fungicide use.
5. **Impact on Food Security:** With a projected 50 percent increase in food demand by 2050, protecting crops from diseases is non-negotiable. Losses of up to 30 percent in staple grains each year directly threaten food availability, prices, and trade flows. Strengthening plant health systems is therefore pivotal to global food security strategies.

6. **Climate Change and Emerging Threats:** Shifting climates alter pathogen lifecycles and geographic ranges, leading to emerging disease pressures in new areas. Studies project that warming trends will expand the habitats of key pathogens, necessitating dynamic monitoring networks and adaptable management practices.
7. **Sustainability and Environmental Stewardship:** Beyond yield protection, plant pathology champions practices that sustain soil health, conserve biodiversity, and minimize agrochemical footprints. By integrating IDM and breeding for resistance, the discipline fosters agricultural systems that are both productive and environmentally responsible.

Importance of Plant Pathology

Plant pathology is far more than an academic subject it underpins the very foundations of agriculture, food security, and ecosystem health. By diagnosing plant diseases early, uncovering their causes, and crafting sustainable solutions, plant pathologists protect harvests valued in the hundreds of billions of dollars and safeguard the world's food supply against a burgeoning population and a changing climate. Their work also preserves biodiversity, supports trade, and enables farmers to adopt environmentally friendly practices that sustain both crops and communities.

1. **Ensuring Food Security:** Plant diseases reduce global food production by an estimated 20–40% each year, creating annual losses exceeding USD 220 billion and threatening the diets of billions of people. In regions like sub-Saharan Africa and South Asia, staple crops such as rice and maize can lose up to 30% of their yield to disease, directly impacting local food availability and nutrition.
2. **Minimizing Economic Losses:** In the United States alone, plant pathogens account for roughly USD 21 billion in crop losses each year, affecting both smallholder farms and large agricultural enterprises. On a global scale, the cumulative economic impact of plant diseases and pests undermines farmers' livelihoods, drives up food prices, and imposes additional costs on consumers and governments alike.
3. **Protecting Ecosystem Services:** Beyond crop yields, plant pathology preserves the ecosystem services that healthy plants provide carbon sequestration, soil stabilization, and habitat for wildlife. Diseases that decimate forests or grasslands compromise these services, with ripple effects on water cycles and biodiversity.

4. **Guiding Sustainable Agriculture:** By developing Integrated Disease Management (IDM) approaches combining biological controls, minimal chemical use, and cultural practices plant pathologists help farmers reduce pesticide dependency and enhance environmental stewardship. IDM frameworks have been shown to lower chemical inputs by up to 50% without reducing yields.
5. **Enabling Trade and Market Access:** Many countries impose strict quarantine regulations to prevent the spread of plant diseases via imported produce. Effective disease surveillance and management allow exporting nations to meet these standards, avoiding trade bans that can cost millions in lost revenue. For example, managing anthracnose and powdery mildew in mangoes has been key to maintaining India's share of global mango exports.
6. **Advancing Breeding and Biotechnology:** Identifying resistance genes in crops is a cornerstone of sustainable disease control. Marker-assisted breeding and genetic engineering have produced varieties resistant to rusts, blights, and other major threats, reducing fungicide applications by up to 70% in some cases.
7. **Responding to Climate Change:** Shifts in temperature and precipitation patterns are expanding the geographic range of many pathogens. Plant pathology research informs predictive models that alert farmers to emerging threats, enabling pre-emptive measures and safeguarding crops against novel disease pressures.
8. **Supporting Public Health:** Some plant pathogens produce toxins that can contaminate food supplies, posing direct risks to human health. Monitoring and controlling these pathogens not only protect crops but also reduce incidences of foodborne illnesses.

The Role of Plant Pathology in Sustainable Agriculture

In an era of climate change and increasing population, plant pathology plays a crucial role in sustainable agriculture. It enables the development of disease-resistant crop varieties, environmentally friendly disease management techniques, and efficient farming practices that reduce crop losses.

India, with its diverse climate and cropping systems, faces numerous challenges related to plant diseases. Therefore, plant pathologists collaborate with agronomists, breeders, and farmers to implement integrated disease management (IDM) practices that combine biological, chemical, and cultural control methods.

Conclusion

Plant pathology is fundamental to the advancement of agriculture. With the rising threat of climate change, emerging pathogens, and increasing demand for food, the role of plant pathology in safeguarding crop health and ensuring food security has never been more critical. Through a combination of scientific research and practical application, plant pathologists continue to contribute to global efforts to meet the agricultural needs of an ever-growing population.

2

Concept of Disease in Plants

Abstract

This chapter provides a comprehensive understanding of the concept of disease in plants, focusing on its definition, causes, symptoms, and wide-ranging significance. A plant disease is defined as a harmful deviation from normal physiological functioning, caused by a continuous interaction between the host, pathogen, and environment—known as the disease triangle. The chapter draws from authoritative sources such as George N. Agrios to explore how diseases disrupt plant structure and function, leading to physiological stress and economic loss. It delves into the visual symptoms of plant diseases—classified into necrotic, chlorotic, hypertrophic, and vascular categories—illustrating each with common examples like leaf spots, blights, root rot, chlorosis, galls, wilting, and vascular discoloration. Beyond the biology, the chapter emphasizes the economic, environmental, social, and political consequences of plant diseases. Case studies, including the Irish Potato Famine, Coffee Rust in Sri Lanka, and Rice Blast in India, underscore the profound impacts plant diseases can have on human societies, agriculture, trade, and food security. By understanding the nature and implications of plant diseases, readers gain insight into the urgent need for effective plant protection strategies.

***Keywords**: Plant disease, disease triangle, symptoms of plant disease, food security,*

Definition and Significance of Plant Diseases

Definition

A plant disease is a harmful deviation from normal functioning caused by a persistent or semi-persistent agent leading to physiological disruptions, structural abnormalities, or economic loss in plants. It manifests through the interaction between the host (plant), pathogen (disease-causing agent), and environment, which together form the **disease triangle**.

According to **George N. Agrios (2005),** a plant disease can be defined as: ***"Any harmful deviation from the normal functioning of physiological processes of plants, induced by continuous irritation of a primary causal agent, leading to symptoms."***

Revised Significance of Plant Diseases

1. **Economic Importance:** Plant diseases lead to large-scale losses in crop production, estimated to account for approximately 10–15% of global crop yields annually (Agrios, 2005). This translates to billions of dollars in losses each year.
2. **Impact on Food Security:** The destruction of staple food crops like rice, wheat, and maize by diseases threatens food security, particularly in developing countries.
3. **Environmental Effects:** Plant diseases contribute to ecosystem imbalances by disrupting natural plant populations and affecting biodiversity.
4. **Social and Political Implications:** Historic plant disease outbreaks, such as the Irish Potato Famine caused by *Phytophthora infestans*, the Bengal Famine aggravated by brown spot disease of rice, and the Coffee Rust Epidemic in Sri Lanka, have led to large-scale human suffering, migration, and social upheaval. These events highlight the interconnectedness between plant health and human societies, with repercussions extending to food security, employment, and political stability.

Differences Between Healthy and Diseased Plants

Healthy Plants

- Exhibit normal growth and development with no signs of physical or physiological stress.
- Perform essential functions like photosynthesis, transpiration, nutrient uptake, and respiration efficiently.
- Remain robust, with high vigour and optimal productivity.

Diseased Plants

- Display disruptions in growth and metabolic processes, leading to abnormal appearances.
- Show symptoms like stunting, yellowing (chlorosis), wilting, or necrosis (death of tissues).
- Suffer from decreased photosynthetic efficiency, nutrient absorption issues, and other physiological stresses, leading to lower yields and poor-quality produce.

Common Symptoms of Plant Diseases

Symptoms are observable signs or manifestations of plant disease, indicating the presence of pathogens that disrupt the plant's normal physiological processes.

These symptoms can vary widely depending on the type of pathogen involved, such as fungi, bacteria, viruses, or nematodes. According to Agrios (2005), plant disease symptoms are generally classified into several broad categories. Here is a detailed explanation:

Necrotic Symptoms

Necrotic symptoms are characterized by the death of plant tissues, leading to visible damage or decay in specific parts of the plant. These include:

a) **Leaf spots:** Leaf spots are localized, often circular or irregularly shaped necrotic lesions that appear on leaves. They are usually caused by fungal or bacterial pathogens. The centre of the spot often turns brown or black due to tissue death, while the surrounding area might appear yellow (a condition called a halo), signalling the progression of infection. **Example :** *Alternaria solani* (Early blight of tomato), *Xanthomonas campestris* (Bacterial leaf spot).

b) **Blights:** Blights refer to the sudden, extensive death of tissue over a wide area of the plant, such as leaves, stems, or flowers. This results in the rapid browning and shrivelling of largeportions of the plant. Blights are often caused by fungi, bacteria, or unfavourable environmental conditions like high humidity.
Example : *Phytophthora infestans* (Late blight of potato and tomato), *Bacillus cereus* (Soft rot).

c) **Cankers:** Cankers are sunken, necrotic lesions that develop on stems, branches, or trunks. These lesions disrupt the flow of water and nutrients within the plant, often leading to further tissue death above the cankered area. Cankers are usually caused by fungi or bacterial pathogens and can severely damage trees and shrubs if left untreated. **Example :** *Cytospora chrysosperma* (Cytospora canker in fruit trees), *Nectria galligena* (Canker disease).

d) **Root rot:** Root rot refers to the decay and death of root tissues, which hinders the plant's ability to absorb water and nutrients from the soil. This condition often leads to wilting, stunted growth, and eventual death of the entire plant if not addressed. Root rot is typically caused by waterlogged soil conditions that favor the growth of pathogenic fungi or bacteria. **Example:** *Pythium spp.* (Damping-off disease), *Fusarium spp.* (Fusarium root rot).

Chlorotic Symptoms

Chlorotic symptoms are characterized by the yellowing of plant tissues, particularly leaves, due to the reduced production or destruction of chlorophyll.

Chlorosis typically affects plant health by reducing photosynthesis and weakening the overall vitality of the plant. These include:

a) **Chlorosis:** Chlorosis refers to a condition where leaves lose their healthy green colour, turning yellow either uniformly or in patches. This yellowing happens when plants are unable to produce enough chlorophyll the pigment responsible for photosynthesis. Several factors can trigger chlorosis, including deficiencies in essential nutrients like iron, nitrogen, or magnesium, as well as viral infections or poor soil drainage. Depending on how severe the problem is, the affected leaves may range in colour from pale green to bright yellow.
   Example: Chlorosis caused by iron (Fe) deficiency; yellowing seen in plants infected by Beet yellows virus.

b) **Mosaic and Mottling:** Mosaic and mottling are symptoms that appear as irregular patches or patterns of light and dark green on plant leaves. These patterns are typically caused by viral infections that disrupt the even distribution of chlorophyll, leading to patchy coloration. In addition to discoloration, infected leaves might also show signs of deformation, such as curling or puckering. These symptoms are commonly associated with viral infections like the Tobacco mosaic virus (TMV) and Cucumber mosaic virus (CMV).
   Example: Tobacco mosaic virus (TMV), Cucumber mosaic virus (CMV).

Hypertrophic Symptoms

Hypertrophic symptoms are characterized by abnormal growth of plant tissues, often resulting in galls, tumours, or excessive branching. These symptoms arise when pathogens disrupt the hormonal balance within the plant, leading to uncontrolled cell proliferation. Specific examples include:

a) **Galls and Tumors:** Galls are abnormal swellings that develop on various plant parts such as leaves, stems, and roots. They occur when pathogens induce excessive cell division. For instance, *Agrobacterium tumefaciens* is a bacterium that infects plants through wounds and transfers its T-DNA into the plant's genome, resulting in tumour-like growths known as crown galls. Additionally, root-knot nematodes (*Meloidogyne spp.*) cause galls on roots by releasing signalling molecules that promote cell division and differentiation.

   Example: *Agrobacterium tumefaciens* causes **Crown Gall Disease**.

   Meloidogyne spp. causes **Root Knot Nematode Disease**.

b) **Witches' Broom:** Witches' broom is a plant disorder that results in abnormal, dense clustering of shoots or branches, giving the plant a bushy or tangled look. This unusual growth is often triggered by disruptions in the plant's hormonal balance, commonly caused by phytoplasmas specialized bacteria that inhabit the phloem. One such organism, *Candidatus Phytoplasma asteris*, is known to cause witches' broom symptoms in a variety of host plants. Viruses can also induce similar effects by interfering with the plant's growth regulation, such as the *Bean yellow mosaic virus* (BYMV), which causes stunted, broom-like growth in infected plants.

Examples: *Candidatus Phytoplasma asteris* – Witches' Broom Disease

Bean yellow mosaic virus (BYMV) – Bean Yellow Mosaic Disease

Wilting and Vascular Symptoms

When a plant's vascular system comprising xylem and phloem is compromised by pathogens, it often results in wilting or internal tissue discoloration. These symptoms signal a serious disruption in water and nutrient transport within the plant.

a) **Wilting:** Wilting is typically observed as drooping or sagging leaves and stems, caused by a loss of turgor pressure. This happens when the plant can no longer move water effectively through the xylem, often due to blockages created by pathogens. Fungal infections like *Fusarium oxysporum* invade and clog xylem vessels, leading to a characteristic wilting known as *Fusarium* wilt. Similarly, bacterial pathogens such as *Xanthomonas campestris* damage the vascular tissues and interfere with water flow, causing bacterial wilt.

Examples: *Fusarium oxysporum* – *Fusarium* Wilt

Xanthomonas campestris – Bacterial Wilt

b) **Vascular Discoloration:** This symptom appears as darkened or brown streaks in the vascular tissues of stems, roots, or leaves. It indicates that the plant's internal transport system has been damaged, usually by fungal pathogens. *Verticillium dahliae*, a soil-borne fungus, is a classic example it infects the xylem and causes dark vascular streaks, leading to wilt and decline in crops like cotton and tomato. Another culprit, *Fusarium solani*, can also trigger vascular discoloration and systemic plant decline.

Examples: *Verticillium dahliae* – *Verticillium* Wilt

Fusarium solani – *Fusarium* Wilt

Economic Impact of Plant Diseases

According to the **International Food Policy Research Institute (IFPRI)** and other agricultural economics studies, the impact of plant diseases extends across multiple facets:

Agricultural Losses

- Annual crop losses due to plant diseases are estimated to exceed 15% of global agricultural output.
- Diseases like **stem rust of wheat**, **blast in rice**, and **late blight in potatoes** have caused major economic disruptions in food production.
- The **2019 locust and wheat rust outbreak** in East Africa caused agricultural damage estimated at hundreds of millions of dollars.

Costs of Control

- Farmers spend significant resources on pesticides, fungicides, and disease-resistant seeds.
- Biological control and chemical management contribute to the rising costs of production, reducing profitability.
- Use of resistant cultivars and crop rotation can help mitigate costs but require initial investment in research and implementation.

Trade and Quarantine

- Plant diseases can lead to restrictions on international trade. For instance, the presence of *Fusarium wilt* (Tropical Race 4) in India has affected the country's banana exports, with many importing nations imposing strict quarantine measures to prevent the spread of this pathogen.
- Quarantine regulations in India, enforced by agencies like the National Plant Quarantine Station (NPQS) under the Directorate of Plant Protection, Quarantine & Storage (DPPQS), aim to control the spread of diseases such as *bacterial wilt* and *brown rot*, significantly impacting both the import and export of agricultural commodities. For example, restrictions are placed on the export of potatoes to prevent the spread of *Ralstonia solanacearum*, the causal agent of bacterial wilt, in other countries.

Social and Political Ramifications

- **The Irish Potato Famine (1845–1852):** The Irish Potato Famine, caused by *Phytophthora infestans*, devastated Ireland's potato crops, which were the primary food source for the population. The disease led to massive crop failures, resulting in widespread hunger and starvation.

Approximately one million people died, and another million emigrated, primarily to the United States. The famine decimated the rural population, leading to social upheaval, a decline in the Irish-speaking population, and significant changes in Irish society. The crisis intensified anti-British sentiments, fueling movements for Irish independence and contributing to political unrest. The event also exposed the dangers of monoculture farming and shifted Ireland's agricultural focus.

- **Coffee Rust in Sri Lanka (1869):** Coffee rust (*Hemileia vastatrix*) devastated Sri Lanka's coffee plantations in the late 19th century, leading to the collapse of the coffee industry, which had been the country's economic backbone. The loss of coffee crops caused widespread economic hardship for farmers and workers reliant on the plantations. Many planters faced financial ruin, while laborers lost their livelihoods. The shift from coffee to tea cultivation had a significant societal impact, as large-scale tea plantations replaced coffee estates. This transition altered the social structure, with new labor systems and the influx of workers from different regions. Tea became Sri Lanka's new economic pillar, but the change also contributed to the marginalization of many traditional coffee-growing communities.
- **Rice Blast in India (Late 19th Century to Present):** Rice blast (*Magnaporthe oryzae*) has been a recurring problem for rice farmers in India, particularly in regions like West Bengal, Tamil Nadu, and Bihar. The disease causes significant yield losses, threatening food security in a country where rice is a staple crop. For farmers, especially smallholders, rice blast leads to financial ruin, as they depend heavily on rice cultivation for income. The disease also exacerbates poverty, especially in rural areas, and increases migration to urban centers. Efforts to control rice blast, such as developing resistant varieties and using fungicides, have had mixed results. The ongoing threat of rice blast underscores the vulnerability of farmers to plant diseases, highlighting the need for more sustainable agricultural practices to safeguard food security and livelihoods.

3

Key Terminologies in Plant Pathology

Abstract

Understanding the terminology used in plant pathology is fundamental for students, researchers, and practitioners working in the field of plant health. This chapter provides clear and concise definitions of essential terms that describe plant-pathogen interactions, disease development, disease management strategies, and related microbiological and ecological concepts. Covering terms from "adjuvants" to "virulence," the chapter serves as a foundational glossary to support deeper learning in plant disease diagnosis, epidemiology, and control. It aims to standardize the language and concepts commonly used in plant pathology, enabling better communication and comprehension in academic and practical contexts.

***Keywords**: Plant pathology, pathogen, disease cycle, infection, symptoms, signs*

Understanding the key terms in plant pathology is crucial for grasping the basic concepts and principles of how diseases develop and how they can be managed. This chapter will define and explain the terminology commonly used in plant pathology to facilitate a clearer understanding of the subject.

- **Adjuvants**: Additives mixed with fungicides to enhance their performance. These are usually inactive substances, such as surfactants, that improve how the chemical spreads, sticks, or penetrates plant surfaces by altering surface tension.
- **Antagonism:** A relationship where one microorganism harms or inhibits the growth of another, often through competition or by producing harmful substances.
- **Antagonists:** Organisms that can suppress or interfere with the growth of other microorganisms, often used in biological control to reduce plant diseases.
- **Antibiosis:** A form of microbial warfare where one organism produces a chemical that negatively affects another organism's growth or survival.
- **Appressorium:** A swollen structure at the tip of a fungal germ tube or hypha that helps the fungus stick to and break into the plant's surface.

- **Avirulent (Non-virulent):** Describes a pathogen that cannot cause disease in a particular host plant.
- **Bactericide:** A chemical substance used specifically to kill or inhibit bacteria.
- **Bacteriology:** The scientific study of bacteria, including their role in plant health and disease.
- **Biotroph:** A type of organism that survives and reproduces only by feeding on living plant tissue.
- **Compatibility:** The ability of two or more substances (like pesticides or chemicals) to be mixed or used together without reducing their effectiveness or causing negative reactions.
- **Competition:** When two organisms try to use the same resource like nutrients or space the one that uses it more effectively dominates.
- **Culture:** The act of growing microorganisms on a prepared nutrient medium under controlled conditions for research or diagnosis.
- **Disease Cycle:** The complete series of steps a disease goes through from survival of the pathogen to infection, colonization, reproduction, and spread.
- **Disease Incidence:** The number or proportion of plants in a population showing disease symptoms.
- **Disease Severity:** A measure of how much damage a disease has caused, often expressed as a percentage of affected plant tissue or yield.
- **Disinfectant:** A chemical or physical agent that eliminates pathogens from plant tissues, tools, or surfaces.
- **Disorder:** A plant problem not caused by a pathogen, but by non-living (abiotic) factors such as nutrient imbalance, pollution, or weather extremes.
- **Eradication:** Controlling disease by removing or destroying the pathogen or infected plant parts to prevent further spread.
- **Exclusion:** Preventing the introduction of pathogens into a clean area, often through quarantine or sanitation practices.
- **Fungicide:** A chemical compound that kills or inhibits the growth of fungi.
- **Hyper-parasitism:** When one parasite infects or lives off another parasite. For example, some fungi can parasitize other fungi.

- **Hyperplasia:** Abnormal enlargement of plant tissues or organs caused by an increase in the number of cells.
- **Hypersensitivity:** A rapid and intense reaction of plant cells to infection, where affected cells die quickly to block the pathogen's spread.
- **Hypertrophy:** Abnormal growth of plant tissues due to an increase in cell size, rather than number.
- **Hypha (plural: Hyphae):** A single, thread-like filament of a fungus. Together, many hyphae form the fungal body.
- **Immune:** Completely resistant to a specific pathogen; the organism cannot be infected.
- **Infection:** The successful entry and establishment of a pathogen inside a plant.
- **Infectious Disease:** A disease caused by a pathogen that can spread from one plant to another.
- **Inoculate:** To deliberately introduce a pathogen to a plant to study its effect or for diagnostic purposes.
- **Inoculation:** The process by which a pathogen comes into contact with a potential host plant.
- **Inoculum:** The actual part of the pathogen (like spores or bacteria) that can start an infection.
- **Inoculum Potential:** A combination of the amount of inoculum present and its ability to cause infection.
- **Isolate:** A pure sample of a microorganism obtained from a single spore or infected plant.
- **Isolation:** The process of separating a pathogen from its host and growing it on a nutrient medium for study.
- **Latent Infection:** An infection where the pathogen is present in the plant, but there are no visible symptoms.
- **Microscopic:** So small that it can only be seen with the help of a microscope.
- **Mycelium (plural: Mycelia):** A mass of fungal hyphae that collectively make up the body of a fungus.
- **Mycology:** The branch of biology that deals with the study of fungi.
- **Necrosis:** Death of plant tissues, often seen as brown or black patches on leaves or stems.

- **Necrotroph:** A type of pathogen that kills plant cells and then feeds on the dead tissue.
- **Nematicides:** Chemicals used to kill or control nematodes (microscopic, worm-like pests).
- **Nematology:** The scientific study of nematodes, especially those affecting plants.
- **Non-Infectious Disease:** Plant disorders caused by non-living factors like nutrient deficiency, chemicals, or temperature not by pathogens.
- **Pathogenesis:** The series of events that occur as a pathogen infects and causes disease in a plant.
- **Pathogenicity:** A pathogen's ability to cause disease in a particular host.
- **Penetration:** The initial entry of a pathogen into the plant's tissues.
- **Phyllody:** A condition in which floral parts transform into green, leafy structures due to pathogen interference.
- **Phytotoxicity:** A harmful effect caused to plants by chemicals, such as pesticides, often resulting in burn or stunted growth.
- **Plant Disease:** Any abnormal condition in a plant that harms its growth, structure, or economic value.
- **Primary Infection:** The first infection in a growing season, usually caused by overwintering or over summering pathogens.
- **Primary Inoculum:** The part of the pathogen that survives between growing seasons and causes the first infections.
- **Quarantine:** Regulations that restrict the movement of plants or plant products to prevent the spread of diseases.
- **Resistant:** A plant's ability to slow down or stop the development of a pathogen after infection.
- **Saprophyte:** An organism that lives on and feeds from dead or decaying organic matter.
- **Secondary Infection:** Infection caused by the spread of pathogens during the same season, following a primary infection.
- **Secondary Inoculum:** The spores or other infectious units produced from earlier infections in the same crop cycle.
- **Sign:** The actual presence of the pathogen or its parts (like fungal spores or bacterial ooze) visible on the plant.

- **Spore:** A tiny reproductive unit produced by fungi and some other organisms, often involved in spreading disease.
- **Susceptible:** A plant that lacks defense mechanisms and is easily infected by a specific pathogen.
- **Symptom:** Visible or measurable effects of disease on a plant, like spots, wilting, or discoloration.
- **Syndrome:** A group of symptoms that consistently occur together and characterize a specific plant disease.
- **Tenacity:** The ability of a pesticide or chemical to stick to plant surfaces and resist being washed off.
- **Tolerance:** The plant's capacity to endure a disease with minimal impact on yield or health, even if infection occurs.
- **Transmission:** The spread of a pathogen from one plant to another, often via wind, water, insects, or tools.
- **Vector:** Any living organism (e.g., insects, mites, nematodes) that carries and transmits a pathogen between plants.
- **Virology:** The scientific study of viruses and their interactions with host organisms.
- **Virulence:** The degree to which a pathogen can cause disease; a measure of its aggressiveness.
- **Virulent:** Describes a pathogen that is highly capable of causing serious disease.
- **Virus:** A sub-microscopic infectious agent made of genetic material (DNA or RNA) enclosed in a protein coat. It cannot reproduce without a host.

4

History of Plant Pathology with Special Reference to India

Abstract

The evolution of plant pathology as a scientific discipline is rooted in centuries of observation and inquiry into the causes of plant diseases. This chapter presents a comprehensive historical account of plant pathology, beginning with ancient Indian contributions from sacred texts and early treatises such as the Arthashastra and Vriksha Ayurveda, which reflect rudimentary understanding of plant health and disease. The progression into scientific mycology began with Theophrastus and later advanced through the pioneering efforts of Leeuwenhoek, Micheli, de Bary, and others, culminating in the modern era of disease management and resistance breeding. Parallel developments in plant bacteriology and virology, from the establishment of germ theory by Pasteur and Koch to the molecular elucidation of viruses and bacteria, are also traced. The discovery of phytoplasmas and spiroplasmas further expanded the scope of plant pathology. Special emphasis is given to the Indian context, highlighting early documentation, indigenous knowledge systems, and the contributions of Indian scholars. This historical journey underscores the interdisciplinary nature of plant pathology and its critical role in ensuring agricultural sustainability and food security.

Keywords: *Plant pathology, history, mycology, bacteriology, virology*

Comprehensive History of Plant Pathology

Mycology

1. **Ancient Contributions (1500–500 BC):** Ancient Indian texts like the *Rigveda* and *Atharva Veda* mention plant diseases, indicating early awareness of pathogens affecting crops. Although these texts are religious and medicinal in nature, they reflect early observations of how environmental factors like fungi and pests could impact plant health, laying the foundation for later agricultural practices.
2. **Theophrastus (286 BC):** The Greek philosopher and botanist, Theophrastus, often called the "Father of Botany," made some of the earliest written references to plant diseases in his work *Enquiry into*

Plants. His observations of plant diseases such as rusts, mildews, and galls were foundational, even though he did not recognize them as caused by fungi. His work represented an important step towards formalizing plant pathology.

3. **Kautilya (321–186 BC):** The ancient Indian scholar Kautilya, also known as Chanakya, in his treatise *Arthashastra*, discussed plant diseases and methods of pest control, reflecting the integration of plant health into agricultural management and governance. His work highlighted the recognition of crop loss as an economic concern during ancient times.
4. **Sushruta Samhita (200–500 AD):** Though primarily a medical text, the *Sushruta Samhita* included references to agricultural practices and plant diseases, showing an understanding that plant health was essential for sustaining human life. This text reflects the broader scope of health, covering not just human but also environmental and agricultural well-being.
5. **Surapal (11th Century):** Surapal, an Indian scholar, authored *Vraksha Ayurveda*, which systematically categorized plant diseases as internal (due to pests or infections) and external (due to environmental conditions). His text was a significant contribution to early plant pathology, emphasizing plant care, protection, and remedies.

Key Developments in Mycology

6. **Anton von Leeuwenhoek (1675):** Leeuwenhoek developed the first microscope, which allowed scientists to observe microorganisms, including fungi. This breakthrough was critical for the future study of fungal pathogens and their impact on plants.
7. **Pier Antonio Micheli (1729):** Italian botanist Micheli proposed that fungi originated from spores. His work, *Nova Plantarum Genera*, marked the beginning of scientific mycology and earned him the title of “Father of Mycology.” He was one of the first to demonstrate the germination of fungal spores under controlled conditions.
8. **Tillet (1755):** French botanist Tillet published findings on “bunt” or “stinking smut” of wheat. His work was the first to recognize a specific fungal disease of crops, suggesting a link between diseases and microorganisms even though the germ theory was not yet fully understood.
9. **I. B. Prevost (1807):** Prevost demonstrated that “bunt” was caused by a fungus and that soaking seeds in a chemical solution could prevent infection. His work laid the groundwork for fungicide use in agriculture.

10. **E. M. Fries (1821):** Fries, a Swedish mycologist, published *Systema Mycologicum*, where he classified thousands of fungal species. His work in fungal taxonomy earned him the title "Linnaeus of Mycology," as it provided a structured way to name and categorize fungi.
11. **The Irish Potato Famine (1845):** The Irish Potato Famine was caused by the fungus *Phytophthora infestans*, leading to widespread crop failure and starvation. This disaster highlighted the severe impact fungal pathogens could have on human societies, prompting further research into plant diseases.
12. **Anton de Bary (1861):** De Bary elucidated the life cycle of *Phytophthora infestans*, proving it to be the cause of potato blight. His work marked a major turning point in plant pathology by showing that fungi could cause disease, rather than merely being the result of unhealthy plants.
13. **H.M. Ward (1881):** Ward worked extensively on coffee rust, earning the title "Father of Tropical Plant Pathology." He demonstrated the life cycle of coffee rust, showing how environmental factors influence the spread and severity of the disease.
14. **Pierre Marie Alexis Millardet (1885):** Millardet discovered the Bordeaux mixture (copper sulfate and lime) by accident while trying to combat downy mildew in grapes. This was the first chemical fungicide used widely in agriculture, revolutionizing fungal disease management.
15. **Robert Hartig (1874, 1882):** Hartig is known as the "Father of Forest Pathology" for his work on diseases of forest trees. His publications on tree diseases and wood decay were pioneering in understanding how fungal pathogens affect forest ecosystems.
16. **R. H. Biffen (1904):** Biffen demonstrated that resistance to fungal pathogens in crops like wheat could be inherited, paving the way for breeding disease-resistant varieties. This discovery connected Mendelian genetics with plant pathology.
17. **Alexander Fleming (1929):** Fleming discovered the antibiotic penicillin from the fungus *Penicillium notatum*. This groundbreaking discovery not only had a major impact on medicine but also highlighted the potential of fungi as sources of biologically active compounds.
18. **H. H. Flor (1942):** Flor proposed the "gene-for-gene" hypothesis in flax rust, suggesting that for every plant resistance gene, there is a corresponding gene in the pathogen that determines its virulence. This laid the foundation for modern plant resistance breeding programs.

Plant Bacteriology

1. **Anton von Leeuwenhoek (1683):** Leeuwenhoek's observations of bacteria through a microscope marked the beginning of bacteriology. His discovery paved the way for the study of bacteria as potential plant pathogens.
2. **Louis Pasteur & Robert Koch (1876):** Pasteur and Koch established the germ theory, proving that diseases in animals and plants could be caused by microorganisms. Koch's postulates became the standard method to prove that a specific pathogen causes a disease.
3. **T. W. D. Sanders (1886):** Sanders described *Bacillus amylovorus* (now *Erwinia amylovora*), the bacterium that causes fire blight in apples and pears. His discovery was crucial in understanding bacterial plant diseases.
4. **Xanthomonas campestris (1887):** The discovery of this bacterium as the cause of black rot in crucifers highlighted the economic impact bacterial diseases can have on agriculture. It led to further research into bacterial pathogens affecting major crops.
5. **L. H. Lentz (1943):** Lentz's work on *Erwinia amylovora* detailed its pathogenic mechanisms, showing how bacterial toxins contribute to fire blight, a serious disease in pome fruit.
6. **Molecular Biology (1980s):** Advances in molecular techniques revolutionized the study of plant bacteriology, allowing for the genetic manipulation of bacteria and the understanding of virulence factors, paving the way for modern bacterial disease management strategies.

Plant Virology

1. **D. Ivanovsky (1892):** Ivanovsky discovered the first plant virus while studying tobacco mosaic disease. He showed that the infectious agent could pass through filters that trapped bacteria, thus identifying viruses as new types of pathogens.
2. **Martinus Beijerinck (1898):** Beijerinck coined the term "virus" and proved that the agent causing tobacco mosaic disease was not a bacterium but a smaller, infectious entity. This finding marked the birth of plant virology.
3. **Tobacco Mosaic Virus (1939):** TMV became the first virus to be crystallized, allowing researchers to study its structure. This milestone advanced the understanding of viral morphology and behavior.

4. **Molecular Virology (1960s):** The development of molecular techniques enabled scientists to analyze viral genomes, leading to the identification of the genes involved in viral replication and plant defense.
5. **RNA Silencing (1990s):** Research into RNA silencing showed how plants defend themselves against viral infections by degrading viral RNA, leading to new strategies for developing virus-resistant crops.

Phytoplasma and Spiroplasma

1. **Doi et al. (1967):** Doi and colleagues first identified phytoplasmas, small bacteria-like organisms that cause diseases like aster yellows. Their discovery initiated research into these unique plant pathogens.
2. **Phytoplasma Research (1970s):** Phytoplasmas were linked to a number of serious plant diseases such as witches' broom and coconut lethal yellowing, increasing awareness of their economic importance.
3. **Molecular Techniques (1990s):** Advances in DNA-based diagnostic tools greatly improved the detection and classification of phytoplasmas and spiroplasmas, aiding in disease management.
4. **Ecological Studies (2010s):** Research into the interactions between phytoplasmas, their insect vectors, and plant hosts led to the development of integrated pest management strategies to control these pathogens more effectively.

5

Causes of Plant Diseases

Abstract

Plant diseases are a major constraint to global agricultural productivity and food security, arising from a complex interplay of non-living (abiotic) and living (biotic) factors. This chapter comprehensively explores the diverse causes of plant diseases. Inanimate causes include adverse environmental conditions—such as extremes in temperature, moisture, and light—as well as nutrient deficiencies and chemical injuries that predispose plants to disease. Detailed insights into how deficiencies in essential nutrients like nitrogen, phosphorus, potassium, calcium, and micronutrients compromise plant physiology and immunity are discussed. Animate causes, on the other hand, involve a wide range of pathogens, including fungi, bacteria, viruses, viroids, nematodes, and protozoa. Each group's mode of infection, disease mechanisms, and common disease examples are systematically covered. Understanding the causes of plant diseases is crucial for diagnosing problems in the field and for designing integrated disease management strategies. This foundational knowledge enables effective prevention and control of plant diseases, contributing to sustainable agricultural practices.

Keywords: *Plant diseases, abiotic stress, biotic stress, environmental factors, nutrient deficiency*

Introduction

Plant diseases significantly impact agricultural productivity and food security, causing substantial economic losses worldwide. The causes of these diseases can be broadly categorized into inanimate and animate factors, each contributing uniquely to plant health. Understanding these causes is essential for developing effective management strategies to prevent and mitigate disease outbreaks.

A. Inanimate Causes

a) **Environmental Factors**: Environmental conditions play a crucial role in the development and severity of plant diseases. Key environmental factors include:

- **Temperature:** Each plant species has an optimal temperature range for growth. Deviations from this range can stress plants, making them more susceptible to pathogens. For example, high temperatures can accelerate the life cycle of some fungal pathogens, while low temperatures can inhibit plant growth, creating conditions conducive to disease.
- **Moisture:** Water is essential for plant growth, but too much of it can become a problem. High humidity levels and frequent rainfall create an ideal environment for many fungal diseases. Pathogens like *Phytophthora* and *Fusarium* thrive under such wet conditions, spreading rapidly in overly moist soils or plant canopies. On the flip side, drought can also lead to trouble. When plants are under water stress, their natural defenses weaken, making them easier targets for both disease-causing organisms (biotic stress) and harmful environmental factors (abiotic stress).
- **Light:** Light plays a vital role in keeping plants healthy. Without enough sunlight, plants can't perform photosynthesis effectively, which limits their energy production and weakens their growth. Plants grown in shaded or low-light conditions often become elongated and pale (a condition called etiolation), and their natural resistance to diseases drops. Inadequate light can also impair the production of compounds involved in plant defense, increasing their vulnerability to pathogens.

b) Nutritional Deficiencies

- **Nitrogen (N) Deficiency:** Nitrogen is crucial for the production of chlorophyll (which makes plants green) and amino acids (the building blocks of proteins). When plants don't get enough nitrogen, their older leaves turn yellow (chlorosis), and overall growth becomes stunted. A reduced leaf area limits photosynthesis, leaving the plant with less energy to defend itself. This weakened state makes it easier for diseases whether fungal, bacterial, or viral to take hold and spread.
- **Phosphorus (P) Deficiency:** Phosphorus plays a key role in energy transfer (as part of ATP) and is especially important for strong root development and flowering. When a plant lacks phosphorus, you might see older leaves turn a dark green or even purplish colour. Roots may not grow well, and the plant matures more slowly. A poorly developed root system makes plants much more vulnerable to soil-borne diseases, such as damping-off or root rots caused by pathogens like *Phytophthora*.

- **Potassium (K):** Potassium regulates water and nutrient transport, as well as the synthesis of proteins and carbohydrates. **Deficiency** in potassium often leads to leaf scorching or yellowing at the leaf margins (marginal chlorosis), curling of leaf tips, and weak stems. Potassium also plays a significant role in enhancing plant disease resistance. Its deficiency compromises the plant's ability to maintain turgor pressure, making the cell walls weaker and more vulnerable to infections from fungal and bacterial pathogens, such as powdery mildew and bacterial blight.
- **Calcium (Ca):** Calcium is fundamental for cell wall structure and stability. **Calcium deficiency** can result in disorders such as blossom end rot in tomatoes, bitter pit in apples, and tip burn in lettuce. Deficiency weakens cell walls, making plants more prone to soft rots and other bacterial and fungal infections. Calcium-deficient plants often display deformed, necrotic leaves, and growth tips may die back due to poor cell division and membrane integrity.
- **Magnesium (Mg):** Magnesium is the central atom in the chlorophyll molecule and is vital for photosynthesis. **Deficiency** typically shows as interveinal chlorosis on older leaves, where the areas between veins turn yellow while the veins remain green. This reduces photosynthetic efficiency, resulting in slower growth and poor energy generation for defense against pathogens. Magnesium-deficient plants are generally weaker and more susceptible to diseases, such as leaf spots and blights.
- **Sulfur (S):** Sulfur is important for protein synthesis and the formation of certain amino acids. **Sulfur deficiency** symptoms resemble nitrogen deficiency, with yellowing of younger leaves and stunted growth. Sulfur-deficient plants may struggle to produce adequate amounts of disease-fighting enzymes and compounds, making them more vulnerable to infections like powdery mildew and fungal rots.
- **Iron (Fe):** Iron is necessary for chlorophyll synthesis and enzyme function. **Iron deficiency** leads to interveinal chlorosis, where young leaves turn yellow while the veins stay green, especially in alkaline soils. Plants deficient in iron are weaker and exhibit reduced metabolic activity, making them more prone to diseases, particularly those caused by nutrient stress like iron-induced chlorosis in citrus.
- **Zinc (Zn):** Zinc is crucial for the production of enzymes and proteins. **Zinc deficiency** results in stunted growth, small and distorted leaves, and shortened internodes. This condition can severely impact plant

vigour, making plants susceptible to infections by bacterial and viral pathogens. Zinc-deficient plants may also have poor seed production, which can limit the plant's ability to recover from disease damage.

- **Manganese (Mn):** Manganese is essential for enzyme activation, photosynthesis, and nitrogen metabolism. **Manganese deficiency** causes interveinal chlorosis, primarily on younger leaves, and can lead to poor plant development. Deficient plants are prone to diseases like leaf spots and fungal infections, as manganese helps protect against oxidative stress induced by pathogen attacks.
- **Copper (Cu):** Copper is vital for lignin formation, which strengthens cell walls, and for electron transport in photosynthesis. **Copper deficiency** results in reduced growth, wilting, and leaf distortion. Plants with copper deficiency have weaker structural integrity, making them highly susceptible to fungal and bacterial infections such as leaf spot, blight, and cankers.

c) **Chemical Injury:** Exposure to harmful chemicals, including pesticides, herbicides, and environmental pollutants, can damage plant tissues and disrupt physiological processes. Chemical injury can result in symptoms such as chlorosis, necrosis, and stunted growth, which can weaken plants and make them more susceptible to pathogen invasion. For instance, excessive use of herbicides can injure non-target plants, leading to increased vulnerability to fungal and bacterial infections.

B. Animate Causes

The animate causes of plant diseases primarily stem from biotic factors, which include pathogens such as fungi, bacteria, viruses, viroids, nematodes, and protozoa.

a) **Fungi:** Fungi are among the most significant plant pathogens, capable of causing a wide range of diseases that can severely impact crop yields. They reproduce through spores, which can be dispersed by wind, water, or insects. Key characteristics of fungi include:

- **Pathogenic Mechanisms:** Fungi can invade plant tissues through natural openings or wounds. They often secrete enzymes that break down plant cell walls, facilitating nutrient absorption. Diseases such as downy mildew (*Peronospora* spp.) and powdery mildew (*Erysiphe* spp.) exemplify the destructive potential of fungal pathogens.

Major Fungal Diseases: Common fungal diseases include rusts, blights, and rots. For instance, *Fusarium* wilt causes vascular wilting in crops like tomatoes and cotton, while *Botrytis cinerea* leads to grey mold in various fruits and vegetables.

b) **Bacteria:** Bacterial pathogens can cause a range of diseases characterized by distinct symptoms. Key aspects include:

- **Morphology and Lifecycle:** Phytopathogenic bacteria are typically rod-shaped or spherical and can multiply rapidly under favourable conditions. They often enter plants through wounds or natural openings, such as stomata.

Common Bacterial Diseases: Notable bacterial diseases include bacterial wilt caused by *Ralstonia solanacearum*, soft rot by *Erwinia* spp., and crown gall induced by *Agrobacterium tumefaciens*. Symptoms may include wilting, rotting, and galls on plant tissues.

c) **Viruses:** Viruses are obligate pathogens that require living host cells for replication. They are often transmitted by insect vectors, mechanical means, or through infected seeds. Key characteristics include:

- **Structure:** Viruses consist of nucleic acid (either DNA or RNA) enclosed in a protein coat. Their small size allows them to invade host cells and hijack the host's machinery for replication.
- **Major Viral Diseases:** Viral infections can lead to stunted growth, yellowing, and mosaic patterns on leaves. Diseases such as tobacco mosaic virus (TMV) and cucumber mosaic virus (CMV) are well-known examples, causing significant yield losses in crops.

d) **Viroids:** Viroids are smaller than viruses and consist solely of a short strand of circular RNA. They are known to cause diseases in certain plant species, primarily affecting flowering plants.

- **Pathogenic Mechanisms:** Viroids disrupt the normal functioning of host cells by interfering with gene expression, leading to stunted growth and other symptoms.
- **Major Viroid Diseases:** The potato spindle tuber viroid (PSTVd) is a significant viroid affecting potatoes and other solanaceous crops, resulting in serious economic losses.

e) **Nematodes:** Plant-parasitic nematodes are microscopic, worm-like organisms that can damage roots and affect plant health. Their feeding activities can lead to nutrient deficiencies and increased susceptibility to other pathogens.

- Pathogenic Mechanisms: Nematodes penetrate plant roots and can cause galls, cysts, or lesions, impacting nutrient and water uptake.
- Common Nematode Diseases: Notable examples include root-knot nematodes (*Meloidogyne* spp.) and cyst nematodes (*Heterodera* spp.), both of which can severely reduce crop yields.

f) **Protozoa:** Protozoa are single-celled organisms that can also act as plant pathogens, though they are less commonly discussed in the context of plant diseases.

- Pathogenic Mechanisms: Some protozoan pathogens, such as *Phytomonas* spp., infect plants and can be transmitted by insect vectors. They can affect various plant parts, leading to symptoms like wilting and stunting.
- Examples of Protozoan Diseases: While protozoan infections are not as prevalent as fungal or bacterial diseases, they can be significant in specific contexts, particularly in tropical and subtropical regions.

6

Classification of Plant Diseases

Abstract

The classification of plant diseases is fundamental to understanding their causes, spread, and management. This chapter presents a comprehensive overview of the various bases for classifying plant diseases, including causal organisms (such as fungi, bacteria, viruses, nematodes, and protozoa), visible symptoms (like leaf spots, wilts, and rots), host range (monocyclic and polycyclic diseases), crop specificity (cereals, legumes, vegetables, fruits), and the infectious or non-infectious nature of diseases. Such a structured classification aids researchers, extension workers, and farmers in accurately diagnosing and managing plant diseases, ultimately contributing to better plant health and agricultural productivity.

Keywords: *Plant disease classification, causal organisms, symptoms, host range,*

Introduction

Plant diseases pose significant challenges in agriculture, affecting crop yields, quality, and overall plant health. To manage these diseases effectively, it is essential to classify them systematically. Accurate classification helps researchers and farmers understand the nature of the diseases, identify the causal agents, and implement appropriate control measures. This chapter explores various ways to classify plant diseases based on different criteria, such as the causal organism, symptoms, host range, type of crops infected, and whether the diseases are infectious or non-infectious.

Basis of Classification

On the basis of various criteria, plant diseases can be classified into the following categories:

1. **Causal Organism**
2. **Symptoms**
3. **Host Range**
4. **Crop Plants Infected**
5. **Infectious vs. Non-infectious Diseases**

A. Causal Organism

Plant diseases are first and foremost classified based on the type of organism responsible for causing the disease. The major causal agents include:

a) **Fungi**: Fungi are among the most prevalent pathogens in plants. They cause diseases leading to symptoms like wilting, blight, and tissue decay. Fungal diseases can be devastating and are responsible for significant crop losses worldwide. Examples include:

   Powdery Mildew *(Erysiphe* spp.*)*: Characterized by white, powdery patches on leaves, stems, and flowers, this disease affects a wide range of plants.

   Rust *(Puccinia* spp.*)*: Causes reddish-brown pustules on leaves, commonly affecting cereals like wheat and barley.

b) **Bacteria**: Bacterial pathogens often result in wilting, leaf spots, and soft rots. They can spread rapidly under favorable conditions, especially in warm and moist environments. Examples of bacterial diseases include:

   Bacterial Wilt *(Ralstonia solanacearum)*: Affects crops such as tomatoes and potatoes, causing wilting and plant death.

   Fire Blight *(Erwinia amylovora)*: Infects apples and pears, causing blackened, scorched-looking shoots and blossoms.

c) **Viruses**: Viral pathogens can cause symptoms such as stunting, leaf yellowing, and mosaic patterns. Viruses can spread through insect vectors, infected seeds, or vegetative propagation. Examples include:

   Tobacco Mosaic Virus (TMV): Causes mottling and yellowing of leaves in tobacco and other crops like tomatoes.

   Cucumber Mosaic Virus (CMV): Leads to leaf yellowing, stunting, and reduced yields in cucumbers and other vegetables.

d) **Nematodes**: Nematodes are microscopic roundworms that attack plant roots, leading to wilting, stunted growth, and poor nutrient uptake. Nematode infestations often go unnoticed until significant damage has occurred. An example is:

   Root-Knot Nematode *(Meloidogyne* spp.*)*: Causes galls (swellings) on plant roots, affecting crops like tomatoes, carrots, and cotton.

e) **Protozoa**: Protozoan diseases in plants are relatively rare but can be problematic in certain crops. They can cause symptoms similar to those caused by fungal or bacterial pathogens.

B. Symptoms

Plant diseases are also classified based on the symptoms they induce. Symptoms can provide clues to the type of disease affecting the plant and guide diagnosis. Common categories include:

a) **Foliar Diseases**: These diseases affect the leaves of the plant and lead to symptoms like:

   Leaf Spots: Small dark lesions or discolored areas on leaves, often caused by fungal or bacterial infections.

   Blight: Rapid and widespread death of plant tissues, leading to brown or blackened leaves. Example: *Early Blight of Tomato (Alternaria solani).*

b) **Vascular Diseases**: These diseases affect the plant's vascular system (xylem and phloem), inhibiting the transport of water and nutrients. Symptoms include:

   Wilting: Loss of turgor in leaves and stems due to impaired water transport.

   Yellowing (Chlorosis): Caused by the breakdown of chlorophyll due to restricted nutrient flow. Example: *Verticillium Wilt (Verticillium spp.).*

c) **Root Diseases**: Root diseases primarily target the root system of plants, often leading to:

Root Rot: Decay and browning of roots, typically caused by soilborne fungi or bacteria.

Stunting: Reduced growth due to poor root function and nutrient uptake. Example: *Phytophthora Root Rot.*

C. Host Range

Diseases can be categorized based on the range of host plants they infect. This classification is particularly useful for understanding the disease spread and control strategies across different plant species.

a) **Monocyclic Diseases**: These diseases typically complete only one infection cycle per growing season and tend to infect a single type of host plant or a very limited range of species. An example is Soybean Rust *(Phakopsora pachyrhizi)*: Primarily affects soybean crops.

b) **Polycyclic Diseases**: These diseases can infect multiple plant species and usually have several infection cycles within a growing season. An example is Downy Mildew (*Peronospora* spp.): Affects a wide range of crops, including cucumbers, onions, and grapes.

D. Crop Plants Infected

Plant diseases are further classified based on the type of crop they infect. This classification is critical for disease management in specific crop groups, as it helps in identifying the particular diseases that may pose threats to each crop category.

a) **Cereals**: Cereal crops such as wheat, rice, maize, and barley are susceptible to several major diseases, including: Wheat Rust *(Puccinia spp.)*: A fungal disease that severely affects wheat, reducing yield. Rice Blast *(Magnaporthe oryzae)*: Causes lesions on rice leaves and stems, significantly impacting production.

b) **Legumes**: Leguminous crops like soybeans, chickpeas, and beans are affected by diseases such as: Ascochyta Blight *(Ascochyta rabiei)*: A fungal disease that causes leaf spots and stem lesions in chickpeas. Bean Rust *(Uromyces appendiculatus)*: Forms rust-colored pustules on bean leaves, reducing photosynthesis.

c) **Vegetables**: Vegetables such as tomatoes, cucumbers, and peppers are highly susceptible to diseases like: Late Blight of Potato and Tomato *(Phytophthora infestans)*: Causes brown lesions on leaves and fruit, leading to crop failure. Fusarium Wilt of Cucurbits *(Fusarium oxysporum)*: Leads to wilting and poor growth in cucumbers.

d) **Fruits**: Fruit crops, including apples, grapes, and citrus, are prone to diseases such as: Citrus Canker *(Xanthomonas axonopodis)*: Causes lesions on citrus leaves, stems, and fruit, affecting fruit quality. Apple Scab *(Venturia inaequalis)*: Causes dark, scabby lesions on apples and leaves, leading to reduced yields and marketability.

E. Infectious vs. Non-infectious Diseases

Plant diseases can also be classified based on their ability to spread between plants.

a) **Infectious Diseases**: These diseases are caused by pathogenic organisms such as fungi, bacteria, viruses, and nematodes. They can spread from one plant to another through various vectors, including air, water, insects, and contaminated tools. An example is *Bacterial Wilt*: Caused by *Ralstonia solanacearum*, transmitted through infected soil or water, and can spread to many plant species.

b) **Non-infectious Diseases**: Non-infectious diseases result from abiotic factors such as environmental stresses, nutrient deficiencies, and chemical injuries. These diseases do not spread between plants but can cause significant damage if the environmental conditions persist. An example is *Nutrient Deficiency*: Physiological disorders caused by lack of essential nutrients like nitrogen, potassium, and phosphorus, leading to poor plant growth but not transmissible between plants.

7

Parasitism and Pathogenesis

Abstract

This chapter delves into the fundamental concepts of parasitism and pathogenesis in plant pathology, explaining how various organisms ranging from fungi and bacteria to viruses and nematodes cause disease in plants. It outlines different types of parasites, including obligate and facultative parasites, with detailed subtypes such as facultative saprophytes and biotrophs. The mechanisms of pathogenesis are discussed in stages: infection, colonization, and disease development, with illustrative examples such as Sclerotium rolfsii on groundnut. Furthermore, the chapter explores the complex dynamics of host-pathogen interaction, focusing on pathogen recognition, effector-mediated suppression of plant defenses, and the role of host resistance. This foundational understanding is key to designing effective disease management strategies in agriculture.

***Keywords**: Parasitism, pathogenesis, obligate parasites, facultative parasites*

Introduction

Parasitism is a biological interaction in which one organism, the parasite, lives on or within another organism, the host, causing harm. In the context of plant pathology, parasitism is a crucial concept, as many plant diseases are caused by various parasitic organisms, including fungi, bacteria, viruses, nematodes, and protozoa. Understanding the mechanisms of parasitism and pathogenesis is essential for developing effective strategies to manage plant diseases.

Definition & Concept of Parasitism

Parasitism is defined as a symbiotic relationship where one organism benefits at the expense of another. In plant pathology, this interaction typically involves pathogenic organisms that exploit host plants for nutrients, ultimately leading to disease.

Types of Parasites

Plant pathogens are generally categorized based on their nutritional relationships with their host plants. These relationships determine how the pathogens interact with the host and whether they can survive outside the host environment. There are two primary types of plant parasites:

a) **Obligate Parasites**: Obligate parasites are organisms that depend completely on living host plants to survive and reproduce. They cannot grow or complete their life cycle outside of a living host. Because of this dependency, they have evolved to live in close association with their hosts, often without killing them immediately. Their survival strategy is to extract nutrients while keeping the host alive long enough to support their growth and reproduction.

Key Characteristics

- Require living tissue to survive.
- Typically have a **narrow host range**.
- Often **biotrophic** - they feed on living cells.
- Cannot grow on artificial culture media.
- Usually cause **chronic** infections with less visible tissue destruction.

Examples

- ***Puccinia graminis***: The fungus that causes stem rust in wheat. It produces reddish-brown pustules on stems and leaves, severely impacting grain yield. It requires living wheat tissue to complete its lifecycle.
- ***Plasmopara viticola***: An oomycete causing downy mildew in grapes. It thrives only in living grapevine tissues, leading to yellowing, curling leaves, and significant crop loss.

a) **Facultative Parasites:** Facultative parasites are more versatile. They can live freely in the environment, feeding on dead organic matter, but they can also switch to a parasitic lifestyle when they encounter a suitable host. This flexibility allows them to survive under diverse conditions, making them opportunistic pathogens that can be highly destructive when infection occurs.

Key Characteristics

- Can survive independently or infect living hosts.
- Usually have a **broad host range**.
- Typically, **necrotrophic** - they kill host tissues and feed on the dead matter.
- Can be grown easily on artificial media.
- Often cause **acute**, aggressive diseases with visible tissue damage.

Examples

- ***Botrytis cinerea***: Known for causing grey mold, this fungus infects many plants. It usually lives on decaying matter but attacks living tissue in moist conditions, especially flowers and fruits, causing a fuzzy grey mold.
- ***Rhizoctonia solani***: A soil-borne fungus that lives saprophytically but infects crops like vegetables and cereals. It causes damping-off and root rot, especially in seedlings, often leading to plant death.

  a) **Subtypes of Facultative Parasites:** Facultative parasites can be further divided into two subcategories based on their primary mode of nutrition:

Facultative Saprophytes

These organisms are mainly saprophytic, thriving on dead organic material, but under specific conditions, they become parasitic.
Example: *Sclerotinia sclerotiorum*, the causal agent of white mold in crops like beans and lettuce, is predominantly saprophytic but can infect living plants when conditions are favorable, especially in cool, moist environments.

Facultative Biotrophs

These are organisms that primarily live parasitically but can occasionally survive in the absence of a host by living saprophytically.
Example: *Colletotrichum spp.*, which causes anthracnose in various plants, is generally parasitic, infecting healthy tissues, but can also survive in dead plant debris between growing seasons.

Mechanisms of Pathogenesis

Pathogenesis refers to the process through which a pathogen causes disease in a host plant. This process involves several stages and mechanisms:

a) **Infection:** The initial step in pathogenesis is the infection of the host plant. This can occur through various pathways:

   Wounds and Natural Openings: Pathogens often enter plants through wounds caused by insects, mechanical damage, or environmental factors. For example, *Erwinia amylovora*, the bacterium causing fire blight in apple and pear trees, can enter through flowers or wounds.

   Direct Penetration: Some pathogens can penetrate the plant cuticle and cell walls directly. Fungal pathogens like *Fusarium* species utilize specialized structures called appressoria to exert pressure on plant surfaces and penetrate host tissues.

b) **Colonization:** Once a pathogen enters the host, it begins to colonize the tissues. This involves:

Nutrient Acquisition: Pathogens extract nutrients from host cells. Fungi and bacteria produce enzymes that break down plant cell walls and release sugars and other compounds, allowing the pathogens to grow and reproduce.

Avoiding Host Defenses: Plants have evolved various defense mechanisms, including physical barriers (like thick cell walls) and biochemical responses (like the production of antimicrobial compounds). Successful pathogens have developed strategies to overcome these defenses, such as producing effector proteins that suppress host immune responses.

c) **Disease Development:** As the pathogen continues to colonize the host, disease symptoms begin to appear. Symptoms can vary widely depending on the pathogen and the plant species but generally include:

Wilting: Often caused by vascular pathogens that disrupt water transport, such as *Verticillium* or *Fusarium* species.

Necrosis: Death of plant tissue, which can manifest as spots, blights, or rotting. For example, late blight (*Phytophthora infestans*) in potatoes leads to extensive tissue necrosis.

Stunting: Reduced growth often results from root pathogens or systemic infections affecting the plant's overall health and nutrient uptake.

Host-Pathogen Interaction

The interaction between the host plant and the pathogen is dynamic and complex, characterized by several stages:

a) **Recognition:** The first step in the host-pathogen interaction is recognition, where the host detects the presence of the pathogen. This process involves:

   Pattern Recognition Receptors (PRRs): Plants possess PRRs that identify pathogen-associated molecular patterns (PAMPs) found on pathogens. This recognition triggers innate immune responses, known as PAMP-triggered immunity (PTI).

b) **Infection and Colonization:** Following recognition, the pathogen can evade or suppress the host's defenses and establish an infection. This process involves:

   Effector Proteins: Many pathogens secrete effector proteins that manipulate host cell functions, often suppressing defense mechanisms. For example, *Pseudomonas syringae* produces effectors that inhibit the host immune response.

c) **Disease Development:** As thc pathogen successfully colonizes the host, it continues to propagate, leading to disease development. The outcome of this interaction depends on various factors, including:

Pathogen Virulence: The ability of a pathogen to cause disease. High virulence strains can cause severe damage, while less virulent strains may result in mild symptoms.

Host Resistance: Some plants have evolved resistance genes that can recognize specific pathogen effectors and trigger stronger immune responses, leading to disease resistance.

Conclusion

Understanding parasitism and pathogenesis is crucial for addressing the challenges posed by plant diseases. By unravelling the complex interactions between plants and pathogens, researchers can develop innovative strategies for disease management, including resistant crop varieties, biological control methods, and targeted fungicides. Continued research in this area is essential for enhancing agricultural productivity and ensuring food security in the face of growing threats from plant diseases.

8

Development of Disease in Plants

Abstract

Chapter 8 delves into the intricate process of disease development in plants, emphasizing the interrelationship among the host, pathogen, and environment commonly illustrated as the disease triangle. It outlines the stages of the disease cycle, including inoculation, incubation, symptom development, dissemination, and survival of the pathogen. Furthermore, the chapter highlights the significance of environmental and host-related factors in influencing disease outbreaks, thereby forming the foundation of plant disease epidemiology. Understanding these dynamic interactions is critical for predicting disease occurrence and designing effective and sustainable plant disease management strategies.

Keywords*: Disease triangle, disease cycle, inoculation, incubation, symptom development, pathogen dissemination*

Introduction

The development of disease in plants is a complex process influenced by the interplay between the host plant, the pathogen, and the environment. Understanding the stages and mechanisms involved in disease development is crucial for effective disease management strategies. This chapter will explore the disease triangle, the disease cycle, and the role of environmental factors in disease epidemiology.

The Disease Triangle

The disease triangle is a fundamental concept in plant pathology that illustrates the three essential components for disease development: the host, the pathogen, and the environment. All three elements must be present for a disease to occur.

a) **Host**: The host is the plant species that can support the growth and reproduction of the pathogen. Different plants exhibit varying levels of susceptibility to diseases, influenced by factors such as genetic resistance, nutritional status, and overall health.

b) **Pathogen**: The pathogen is the organism causing the disease, which can be fungi, bacteria, viruses, or other organisms. Each pathogen has specific requirements for infection and disease development, including particular environmental conditions and host factors.

c) **Environment:** Environmental factors play a critical role in disease development. These include:

   i. **Temperature:** Many pathogens have optimal temperature ranges for growth and reproduction. For instance, *Phytophthora infestans*, the causal agent of late blight in potatoes, thrives in cool, moist conditions.

   ii. **Moisture:** High humidity and free water on plant surfaces can promote the germination of spores and enhance the infection process. Rainfall, irrigation practices, and dew formation can significantly impact disease incidence.

   iii. **Light:** Some pathogens require specific light conditions for spore germination or infection. For example, certain fungal pathogens are more active during periods of high light intensity.

The interaction of these three components determines the potential for disease development. If any one of the components is absent or unfavourable, disease is unlikely to occur.

The Disease Cycle

The disease cycle refers to the sequence of events that occur from the initial infection of the host by the pathogen to the eventual development of disease symptoms and the release of new inoculum.

The key stages of the disease cycle include:

a) **Inoculation:** Inoculation is the first step in the disease cycle, where the pathogen is introduced to the host plant. This can occur through various means:

   Airborne Spores: Many fungal pathogens produce spores that are carried by wind or rain to infect distant plants.

   Insect Vectors: Insects can transmit viruses and bacteria from infected plants to healthy ones. For example, aphids are known vectors for several viral diseases.

   Mechanical Transmission: Human activities, such as pruning or harvesting, can inadvertently introduce pathogens to healthy plants.

b) **Incubation:** After inoculation, the pathogen must establish itself within the host. The incubation period is the time between infection and the appearance of symptoms. This period can vary widely depending on factors such as:

Pathogen Type: Different pathogens have different rates of development. For example, fungal pathogens may have shorter incubation periods compared to bacterial pathogens.

Host Condition: A healthy host with good nutritional status may show symptoms later than a stressed plant.

Environmental Conditions: Favourable conditions for pathogen growth can shorten the incubation period. For instance, high humidity may speed up fungal growth.

c) **Symptom Development:** Following the incubation period, symptoms of the disease begin to appear. Symptoms can manifest in various ways, including:

Wilting: Often seen in vascular diseases caused by pathogens like *Fusarium* or *Verticillium*, which block water transport.

Leaf Spots: Many fungal and bacterial diseases lead to localized necrosis, resulting in leaf spots.

Cankers and Galls: Some pathogens induce abnormal growths, such as galls in the case of crown gall disease caused by *Agrobacterium tumefaciens*.

d) **Dissemination:** Once symptoms are visible, the pathogen continues to reproduce and spread. Dissemination can occur through several methods:

Spore Production: Fungi produce vast numbers of spores that can be dispersed by wind, water, or insects.

Mechanical Means: Pathogen-laden plant debris can serve as a source of inoculum, especially if left on the ground after harvest.

Insect Transmission: Insects can carry pathogens from one plant to another, contributing to the spread of the disease.

e) **Survival:** After disease symptoms have faded, pathogens can survive in various forms until they encounter new hosts. Survival mechanisms include:

Dormant Spores: Many fungi and bacteria can form resistant spores that remain viable in the environment for extended periods.

Seed Transmission: Some pathogens can be transmitted through infected seeds, ensuring their survival until the next planting season.

Plant Debris: Pathogens can persist in infected plant tissues, allowing them to infect subsequent crops.

Epidemiology of Plant Diseases

Epidemiology is the study of how diseases spread and the factors influencing their occurrence. Understanding the epidemiology of plant diseases is essential for developing effective management strategies. Key factors influencing disease spread include:

a) **External Factors: Weather Conditions:** Temperature, humidity, and precipitation significantly influence pathogen survival and disease development. For example, warm, wet conditions are ideal for fungal diseases.

   Agricultural Practices: Practices such as crop rotation, planting density, and irrigation can affect disease spread. Continuous cropping of susceptible plants can lead to increased disease pressure.

   Geographical Spread: The distribution of crops and the movement of plant material can influence how diseases spread. For instance, the introduction of new plant varieties may inadvertently introduce pathogens to new areas.

b) **Host Factors: Resistance Levels:** Some plants have evolved resistance mechanisms that reduce the likelihood of disease development. Understanding these resistance traits is crucial for breeding disease-resistant varieties.

Nutritional Status: Well-nourished plants are generally more resistant to diseases than those experiencing nutritional deficiencies.

Conclusion

The development of disease in plants is a multifaceted process influenced by the interaction of the host, pathogen, and environment. Understanding the stages of disease development and the factors that affect disease epidemiology is essential for effective disease management strategies. By addressing these components, researchers and practitioners can enhance crop health and productivity, ultimately contributing to food security and sustainable agriculture.

9

Fungi and Their Role in Plant Pathology

Abstract

Fungi are a diverse group of eukaryotic organisms that play a critical role in plant pathology as major pathogens causing a variety of plant diseases. This chapter explores the general characteristics of fungi, including their unique cell structure, nutritional modes, and reproductive strategies. The morphology of fungi, focusing on hyphae and spores, is detailed with emphasis on both asexual and sexual spores and their roles in fungal propagation and survival. The chapter also explains the formation and types of fungal fruiting bodies across different fungal divisions, highlighting important pathogenic examples. Furthermore, fungal reproduction mechanisms both asexual and sexual are described, providing insights into fungal life cycles and their implications for plant disease epidemiology and management.

Keywords: *Fungi, Plant Pathology, Hyphae, Spores, Asexual Reproduction, Sexual Reproduction*

Introduction

Fungi are a diverse group of eukaryotic organisms that play a significant role in ecosystems, including their function as pathogens in plants. They can cause a wide range of diseases that affect plant health and agricultural productivity. Understanding the characteristics, life cycles, and impact of phytopathogenic fungi is crucial for effective disease management.

General Characteristics of Fungi

Fungi are distinguished from plants and animals by several key features:

Cell Structure: Fungi have a complex cell structure with a rigid cell wall made primarily of chitin, which differs from the cellulose found in plant cell walls.

Nutritional Mode: Fungi are heterotrophic organisms that obtain nutrients through external digestion. They secrete enzymes into their environment to break down complex organic matter into simpler compounds, which are then absorbed.

Reproduction: Fungi can reproduce both sexually and asexually. Asexual reproduction typically occurs through the production of spores, while sexual reproduction involves the formation of specialized structures for mating.

Morphology of Fungi

a) **Hyphae:** Hyphae are the fundamental units of fungal structure. They are long, thread-like filaments that form a network called **mycelium**. Hyphae can be classified into:

   Septate Hyphae: Contain cross-walls (septa) that divide them into individual cells.

   Coenocytic Hyphae: Lacks septa and consists of continuous cytoplasmic mass with multiple nuclei.

b) **Spores:** Spores are the reproductive units of fungi. They can be produced sexually or asexually and are essential for dispersal and survival in adverse conditions. Spores are categorized into:

   Asexual spores play a crucial role in the reproduction and dissemination of many fungi. These spores are produced without the involvement of gametes, allowing fungi to reproduce rapidly under favorable conditions. There are several types of asexual spores, each formed through different mechanisms and on various specialized structures. Here's a detailed description of common asexual spores:

   1. **Arthrospores:** Arthrospores are asexual spores that form when fungal hyphae break into individual cells. Each segment, now separated from the hypha, acts like a spore capable of growing into a new organism. These spores usually have thick walls and help the fungus survive tough conditions rather than spread quickly. They are commonly found in fungi like *Geotrichum*, which can cause rots in fruits and other plant issues.

   2. **Chlamydospores:** Chlamydospores are thick-walled, round or oval spores formed inside the fungal hyphae. They act like a survival capsule, helping the fungus withstand harsh environments such as drought, cold, or nutrient-poor conditions. Unlike spreading spores, chlamydospores mostly wait for better conditions to grow again. Fungi such as *Fusarium* and *Candida* often produce these tough resting spores.

   3. **Blastospores:** Blastospores are formed through budding a process where a small outgrowth develops from a parent fungal cell, grows in size, and then separates to live on its own. This method is typical in yeasts like *Saccharomyces* and *Candida*. Blastospores are suited for

quick multiplication, especially when conditions are favourable for growth and spread.

4. **Sporangiospores:** Sporangiospores are formed inside a balloon-like structure called a sporangium, which sits at the end of a stalk called a sporangiophore. When mature, the sporangium bursts open, releasing the spores into the air, water, or onto nearby surfaces. Fungi like *Rhizopus* and *Mucor*, known for causing bread mold, produce large numbers of these spores to spread quickly and colonize new areas.
5. **Conidiospores (Conidia):** Conidiospores, also called conidia, are produced externally on structures called conidiophores. Unlike sporangiospores, conidia are not enclosed in a sac they're formed openly at the tip or side of the conidiophore and released directly into the surroundings. They can differ in shape and size and are usually produced in high numbers. Fungi like *Aspergillus* and *Penicillium* use conidia to spread rapidly, often becoming major agents in plant diseases due to their efficient dispersal by air or water.
6. **Sexual Spores (Ascospores, Basidiospores):** Result from sexual reproduction; they are formed within specialized structures called asci (in ascomycetes) or basidia (in basidiomycetes).

c) **Fruiting Bodies:** Fruiting bodies are complex, multicellular structures formed by fungi during their reproductive phase. These specialized structures serve as the sites where spores, either sexual or asexual, are produced and released into the environment for dispersal. Fruiting bodies vary greatly in form, size, and complexity, depending on the type of fungus and its reproductive strategy. Fungi are classified into different divisions based on their spore-producing structures and reproductive methods. The following section elaborates on the key asexual and sexual fruiting bodies of fungi, arranged division-wise, with important examples and detailed descriptions of structures like aecium, coremium, sporodochium, puffballs, and bracket fungi.

Division: Ascomycota (Sac Fungi)

Ascomycota is the largest group of fungi and includes many important plant pathogens and beneficial fungi. They reproduce sexually through structures called asci, which produce ascospores, and asexually through conidia.

Asexual Fruiting Bodies

a) **Sporodochium**: A compact, cushion-like mass of conidiophores that bear conidia. These structures are typically found on the surface of plant tissues and help with the rapid spread of conidia to new hosts.

Example: *Fusarium oxysporum*, the causal agent of Fusarium wilt in many crops, produces sporodochia on plant surfaces.

b) **Coremium (Synnemata)**: Conidiophores bundle together to form a stalk-like structure, elevating the conidia for more efficient dispersal.**c. Conidiophores**: Specialized structures on which conidia (asexual spores) are produced, often forming on the ends of hyphae.

Example: *Penicillium* species produce conidiophores with characteristic brush-like structures bearing conidia.

Sexual Fruiting Bodies

a) **Cleistothecia:** Cleistothecia are completely closed, spherical fruiting bodies that enclose **asci** (sac-like structures) containing **ascospores**. Since there's no opening, spores are released only when the cleistothecium breaks open, typically due to environmental changes or decay. **Example:** *Erysiphe* species, which cause powdery mildew, produce cleistothecia. These fungi appear as white, powdery coatings on leaves and stems.

b) **Perithecia:** Perithecia are flask-shaped fruiting bodies with a small opening at the top called an **ostiole**. Inside, they contain asci that release spores through this opening, often with some force. **Example:** *Claviceps purpurea*, the fungus behind ergot disease in cereals, develops perithecia on its hardened fungal structures (sclerotia) during its sexual stage.

c) **Apothecia:** Apothecia are open, cup- or disc-shaped fruiting bodies with exposed asci. This structure makes it easier for ascospores to be dispersed by wind or rain when conditions are favourable.
Example: *Sclerotinia sclerotiorum*, which causes white mold in crops, forms apothecia from its overwintering structures (sclerotia) in cool, moist environments.

d) **Aecium:** An aecium is a type of fruiting body found in **rust fungi**, responsible for producing **aeciospores** during one stage of their complex life cycle. Aecia often appear as clustered, blister-like structures on the underside of leaves. **Example:** *Puccinia graminis*, the wheat stem rust fungus, produces aecia on alternate hosts like barberry as part of its multi-host, multi-stage life cycle.

Division: Basidiomycota (Club Fungi)

Members of the **Basidiomycota** division reproduce sexually by producing **basidiospores** on specialized, club-shaped cells known as

basidia. While asexual reproduction occurs, it is less common in this group.

Sexual Fruiting Body: Mushrooms (Basidiocarps)

Mushrooms are the most recognizable fruiting bodies of Basidiomycota. They typically consist of a **cap** (pileus) and a **stem** (stipe). The underside of the cap has **gills or pores** that are lined with basidia, where spores are produced and released.

Examples:

- *Agaricus bisporus*: The common edible button mushroom.
- *Amanita muscaria*: A toxic mushroom known for its bright red cap with white spots.

e) **Bracket Fungi (Polypores)**: These fungi form woody, shelf-like structures on trees or decaying logs. The basidiospores are produced inside tubes or pores located on the underside of the bracket.

   Example: *Ganoderma lucidum* (Reishi), a medicinal mushroom, and *Fomes fomentarius*, which causes white rot in trees.

f) **Puffballs**: Puffballs are spherical fruiting bodies that release basidiospores in a cloud when the outer surface ruptures due to mechanical impact or pressure.

   Example: *Lycoperdon perlatum*, commonly found in grasslands and forests, releases spores through a pore at the top when disturbed.

Division: Deuteromycota (Imperfect Fungi)

Deuteromycota refers to fungi that reproduce only asexually, as their sexual stage has not been observed or is unknown. Once sexual stages are identified, they are often reclassified into Ascomycota or Basidiomycota.

Asexual Fruiting Bodies

a) **Sporodochium**: A mass of conidiophores that form cushion-like structures to produce conidia. These are found on plant surfaces and can produce massive quantities of spores for rapid disease spread.

   Example: *Fusarium* species, which cause diseases like wilts, blights, and root rots, produce sporodochia.

b) **Coremium (Synnemata)**: These structures consist of aggregated conidiophores that elevate conidia, making dispersal by wind or water more efficient.

   Example: *Graphium* species, involved in wood decay and plant diseases, produce coremia.

Sexual Fruiting Bodies

None: By definition, Deuteromycota does not exhibit sexual reproduction, and no sexual fruiting bodies are formed. However, many of these fungi have been reclassified into Ascomycota or Basidiomycota when their sexual stages are discovered.

Summary of Fruiting Bodies Division-wise

Division	Asexual Fruiting Bodies	Sexual Fruiting Bodies	Examples
Zygomycota	Sporangia: Sac-like structure producing sporangiospores	Zygospores: Thick-walled spores from fusion of hyphae	*Rhizopus stolonifer* (Black Bread Mold), *Mucor* species
Ascomycota	Sporodochium: Cushion-like mass of conidiophores Coremium (Synnemata): Stalk-like conidiophore bundle Conidiophores: Structure bearing conidia	Cleistothecia: Closed spherical structures Perithecia: Flask-shaped with ostiole Apothecia: Cup-shaped open structure Aecium: Produces aeciospores in rust fungi	*Fusarium oxysporum*, *Penicillium* spp., *Claviceps purpurea*, *Erysiphe* spp., *Puccinia graminis*
Basidiomycota	Coremium (Synnemata) (rare)	Mushrooms (Basidiocarps): Cap and stem with gills Bracket Fungi: Shelf-like structures on wood Puffballs: Spherical, releases spores when ruptured	*Agaricus bisporus* (Edible Mushroom), *Ganoderma lucidum* (Reishi), *Lycoperdon perlatum* (Puffball)
Deuteromycota	Sporodochium: Cushion-like mass of conidiophores Coremium (Synnemata): Stalk-like conidiophore bundle	None (No sexual reproduction observed)	*Fusarium* species, *Graphium* species

Reproduction in Fungi

Asexual Reproduction in Fungi

Asexual reproduction in fungi is a highly efficient method for propagating species without the need for genetic recombination. It occurs in both unicellular and multicellular fungi and can take several forms, allowing fungi to colonize new environments rapidly.

Asexual Reproduction in Unicellular Fungi

1. **Fission:** In this process, a parent cell splits into two equal-sized daughter cells. This is typical in unicellular fungi like *Schizosaccharomyces*, a type of yeast. Each daughter cell is genetically identical to the parent and capable of independent growth, contributing to population expansion.
2. **Budding:** This is a common mode of reproduction in yeasts such as *Saccharomyces cerevisiae* (brewer's or baker's yeast). A small outgrowth, or bud, forms on the parent cell, grows in size, and eventually separates to become a new organism. The bud is initially smaller than the parent but matures over time. Budding allows rapid and continuous reproduction, especially in nutrient-rich conditions.

Asexual Reproduction in Multicellular Fungi

1. **Fragmentation:** In this method, the fungal mycelium (the vegetative part of the fungus) breaks into smaller pieces. Each fragment can grow independently into a full mycelium, provided it lands in a suitable environment. This type of reproduction is seen in various molds, where mechanical breakage of hyphae leads to propagation.
2. **Asexual Spore Formation:** The most common form of asexual reproduction in multicellular fungi. Spores are produced by mitosis and are genetically identical to the parent. These spores are usually dispersed by wind, water, or animals and can grow into new mycelia when they land in favorable conditions. Asexual spores can be produced in different ways:
   - **Conidia:** Produced by conidiophores (specialized hyphae), as in *Aspergillus* or *Penicillium*.
   - **Sporangiospores:** Formed within sporangia, which are sac-like structures found in fungi such as *Rhizopus* (bread mold).

Types of Asexual Spores

1. **Sporangiospores**: These are non-motile spores produced in a **sporangium** (a sac-like structure) at the tip of **sporangiophores** (specialized hyphae). Some fungi, like *Rhizopus*, produce **zoospores** (motile spores with flagella), but these are rare in terrestrial fungi.
2. **Conidia**: **Conidia** are non-motile asexual spores produced by fungi like *Penicillium, Aspergillus*, and many other molds. Conidia are produced at the tips or sides of specialized hyphae called **conidiophores**. Conidia are usually dispersed by air and can be produced in large numbers, enabling fungi to spread over large distances.**Blastospores**: Blastospores are

formed by budding from a parent cell or hyphal element. These spores are common in yeasts and some filamentous fungi.

3. **Chlamydospores**: Chlamydospores are thick-walled, resistant spores that are formed in unfavorable conditions. They can survive harsh environmental conditions and are produced terminally or intercalarily in hyphae. Examples: *Candida* species and some members of the *Zygomycetes*.
4. **Oidia (Arthrospores)**: Oidia are thin-walled spores that form when hyphal cells fragment, breaking apart into individual spores. These spores germinate directly into new hyphae under favorable conditions.

Asexual Structures

1. **Sporodochia**: Small, cushion-like masses of conidiophores.
2. **Pycnidia**: Flask-shaped fruiting bodies containing conidia.
3. **Aceruli**: Flat or cushion-shaped fruiting structures where conidiophores arise in compact masses.

Sexual Reproduction in Fungi

Sexual reproduction involves the fusion of two nuclei, which leads to genetic recombination and variability. It occurs through the production of **sexual spores** that result from the fusion of haploid nuclei from opposite mating types. The process involves three key steps: **plasmogamy**, **karyogamy**, and **meiosis**.

Key Stages of Sexual Reproduction:

1. **Plasmogamy**: The fusion of the cytoplasm from two different mating types or gametes. This creates a **dikaryon** (a cell with two distinct nuclei).
2. **Karyogamy**: The fusion of the two nuclei to form a diploid **zygote** nucleus.
3. **Meiosis**: The diploid nucleus undergoes meiosis, resulting in the formation of haploid spores. These spores give rise to genetically diverse offspring.

Types of Sexual Reproduction in Fungi:

1. **Isogamy**: Fusion of similar gametes from two different mating types (e.g., some *Chytridiomycota*).
2. **Anisogamy**: Fusion of dissimilar gametes, where one gamete is larger (e.g., some *Oomycota*).
3. **Oogamy**: A specialized type of anisogamy where a large non-motile egg cell fuses with a smaller motile sperm (e.g., *Saprolegnia*).

4. **Gametangial Contact**: A form of sexual reproduction where male and female reproductive organs (gametangia) come into contact, and male nuclei are transferred to the female organ through a pore. **Oomycetes** (e.g., *Phytophthora*) exhibit this form, producing **oospores** after plasmogamy.
5. **Gametangial Copulation**: Complete fusion of the entire contents of male and female gametangia into one structure, leading to the formation of a **zygospore** (e.g., *Mucor* and *Rhizopus*).
6. **Spermatization**: Male reproductive structures called **spermatia** (e.g., in rust fungi) are transferred to receptive female hyphae. Plasmogamy occurs when the spermatium fuses with the female organ.
7. **Somatogamy**: A process in which sexual reproduction occurs without the involvement of specialized sex organs. Instead, two vegetative hyphae fuse and result in a dikaryotic mycelium (e.g., in mushrooms like *Agaricus*).

Types of Sexual Spores:

1. **Oospores:** Oospores are thick-walled sexual spores formed as a result of sexual reproduction in the *Oomycota* (water molds and downy mildews). These spores are developed within a structure called the oogonium, following fertilization between the male antheridium and the female oogonium. The oospore acts as a survival structure, allowing the organism to withstand unfavorable environmental conditions such as desiccation or freezing. Oospores can germinate when conditions improve, giving rise to a new mycelium. Examples of fungi that produce oospores include *Phytophthora infestans* (causal agent of late blight in potatoes) and *Pythium* species (causal agents of damping-off in seedlings).
2. **Zygospores:** Zygospores are formed by the sexual fusion of two similar or compatible gametangia (sexual reproductive structures) in *Zygomycota* fungi. These spores have a thick, resistant outer wall, which enables them to survive in extreme or unfavorable environmental conditions for long periods. The process of zygospore formation involves the fusion of specialized hyphae called gametangia from two different mating strains of fungi. Zygospores germinate when conditions become favorable, producing sporangia that release haploid spores. They are commonly found in fungi such as *Rhizopus* and *Mucor*, which are responsible for bread mold and fruit rot.
3. **Ascospores:** Ascospores are the result of sexual reproduction in fungi belonging to the *Ascomycota* division, which includes important species

like *Neurospora* (a model organism in genetics) and *Saccharomyces cerevisiae* (baker's yeast). These spores are produced within a sac-like structure called an ascus (plural: asci). Asci are typically found within specialized fruiting bodies called ascocarps, such as cleistothecia, perithecia, or apothecia, depending on the species. After the fusion of compatible hyphae, meiosis occurs in the ascus, followed by mitosis, resulting in eight haploid ascospores. Ascospores are a key feature of this group of fungi, contributing to their dispersal and survival.

4. **Basidiospores:** Basidiospores are sexual spores formed in fungi belonging to the *Basidiomycota* division, which includes familiar mushrooms, puffballs, and rust fungi. These spores are produced on the surface of specialized club-shaped cells called basidia (singular: basidium). Basidia are often found on the gills, pores, or other reproductive surfaces of the fruiting bodies, such as the gills of mushrooms (*Agaricus bisporus*). Following meiosis, basidiospores are typically ejected from the basidium and dispersed by wind. Each basidiospore is haploid, and upon germination, it forms a new mycelium. The process is central to the reproductive cycle of *Basidiomycota* fungi, which includes both plant pathogens and edible fungi.

Examples of Sexual Reproduction in Different Fungal Phyla:

1. **Chytridiomycota** (Chytrids): These fungi have motile zoospores and exhibit sexual reproduction through isogamy or anisogamy.

   Example: *Synchytrium endobioticum*, which causes potato wart disease.
2. **Zygomycota** (Zygomycetes): Sexual reproduction occurs by the formation of **zygospores** after gametangial copulation.

   Example: *Rhizopus stolonifer* (bread mold).
3. **Ascomycota** (Sac Fungi): These fungi form sexual spores called **ascospores** in a sac-like structure known as the **ascus**.

   Example: *Neurospora crassa*, used as a model organism in genetics research.
4. **Basidiomycota** (Club Fungi): These fungi form sexual spores called **basidiospores** on structures known as **basidia**.

   Example: *Agaricus bisporus* (the common mushroom).

Important fungal genera in India

1. ***Fusarium*** *Fusarium* species, notably *Fusarium oxysporum*, are notorious pathogens causing *Fusarium* wilt in various crops such as pigeonpea, cotton, and banana. Morphologically, they are characterized

by their slender, elongated macroconidia, which are sickle-shaped, and smaller microconidia, typically oval to spherical in shape. These fungi invade plant tissues and produce mycotoxins, leading to wilting, yellowing, and eventually death of the affected plants. Fusarium can persist in the soil, making management challenging. Eg. *Fusarium* wilt in cotton and banana.

2. ***Aspergillus***: *Aspergillus flavus* and *Aspergillus niger* are two prominent species that significantly impact crops like groundnut and maize. *Aspergillus flavus* is infamous for producing aflatoxins, highly toxic compounds that can contaminate food and feed, posing severe health risks. Morphologically, *Aspergillus* species are identified by their conidiophores, which bear globose vesicles, and chains of conidia that are usually green or black in color. The environmental conditions favoring their growth, such as high humidity and temperature, increase the risk of aflatoxin contamination. Eg. Aflatoxin contamination in groundnut and maize.
3. ***Alternaria:*** *Alternaria alternata* is a major pathogen responsible for *Alternaria* leaf spot and blight, particularly in crops such as tomato, potato, and mustard. The fungus features dark, round to oval conidia with characteristic longitudinal and transverse septa. It thrives in warm and humid conditions, where its spores can easily spread through rain splashes or irrigation. Infected plants show symptoms of yellowing and leaf drop, which can significantly reduce yield and quality. Eg. *Alternaria* leaf spot in tomato.
4. **Rhizoctonia:** *Rhizoctonia solani* is responsible for sheath blight in rice and root rot in various pulse crops. The fungus is characterized by its irregularly branched, brown mycelium and sclerotia, which can survive in soil for long periods. Infected plants exhibit symptoms like water-soaked lesions on leaves and rotting of roots and stems. Effective management strategies often involve crop rotation and the use of resistant varieties. Eg. Sheath blight in rice.
5. **Sclerotium:** *Sclerotium rolfsii* is a soil-borne pathogen that causes stem rot in groundnut and foot rot in sugar beet. The fungus produces white mycelium and sclerotia that can survive for several years in soil. It typically invades through wounds or at the soil line, leading to symptoms such as yellowing and wilting of plants, followed by decay of the stems. Management strategies include crop rotation and proper soil drainage to minimize moisture levels conducive to infection. Eg. Stem rot in groundnut.

6. ***Colletotrichum***: *Colletotrichum gloeosporioides* is a significant plant pathogen that leads to anthracnose in various fruits, including mango, chili, and papaya. The fungus produces acervuli containing dark conidia, which are disseminated by water splashes. Symptoms of anthracnose include dark, sunken lesions on fruits and leaves, often leading to premature fruit drop and significant economic losses. Effective management practices include regular monitoring and the application of fungicides. Eg. Anthracnose in mango and papaya.
7. ***Puccinia***: *Puccinia triticina* is responsible for wheat leaf rust, a major disease affecting wheat production in India. This rust fungus produces reddish-brown urediniospores on infected wheat leaves. The disease can spread rapidly, particularly during warm and humid weather, leading to significant yield losses. Management strategies often include the use of resistant wheat varieties and timely application of fungicides. Eg. Wheat leaf rust.
8. ***Ustilago***: *Ustilago maydis*, known as corn smut, is prevalent in maize-growing regions. The fungus produces black, powdery teliospores that develop on the ears of maize, leading to galls that are considered a delicacy in some cuisines. While it can be tolerated or even sought after in culinary contexts, it still causes economic losses for maize farmers. Management involves planting resistant varieties and managing field sanitation. Eg. Corn smut in maize.
9. ***Curvularia***: *Curvularia lunata* is a pathogen that causes leaf spot in rice and blight in maize. The fungus is characterized by its curved or crescent-shaped conidia and thrives in warm, humid environments. Symptoms include dark lesions on leaves, which can impair photosynthesis and lead to reduced yields. Effective management includes crop rotation and fungicide applications. Eg. Leaf spot in rice.
10. ***Phytophthora* (Oomycete):** While not a true fungus, *Phytophthora infestans* is a devastating pathogen causing late blight of potato, significantly impacting potato cultivation in India, particularly in states like West Bengal and Punjab. The pathogen produces sporangia and zoospores that facilitate its spread during wet conditions. Symptoms include dark lesions on leaves and tubers, often leading to complete crop loss if not managed effectively. Eg. Late blight in potato.
11. ***Cercospora***: *Cercospora personata* is responsible for leaf spot and blight in groundnut. The fungus produces conidia that are elongated and cylindrical, leading to dark lesions on leaves that can result in defoliation and reduced yields. Effective management practices include the use of

resistant varieties and timely fungicide applications. Eg. Leaf spot in groundnut.

12. ***Botrytis*:** *Botrytis cinerea*, also known as grey mold, affects various crops such as grapes, strawberries, and tomatoes. The fungus produces grey, fluffy conidia on decaying plant tissues and thrives in high humidity conditions. Infected plants exhibit symptoms of wilting and decay, leading to significant post-harvest losses. Management strategies include the application of fungicides and improving air circulation in fields. Eg. Grey mold in grapes.
13. ***Pythium*:** *Pythium* spp. are significant pathogens causing damping-off in seedlings and root rot in various crops. They are characterized by their oospores and filamentous mycelium, thriving in wet conditions that facilitate their spread. Infected seedlings exhibit symptoms of wilting and decay, emphasizing the importance of proper soil drainage and sanitation practices. Eg. Damping-off in various crops.

Summary

Fungi exhibit a diverse range of reproductive strategies, both asexual and sexual, allowing them to thrive in various environments. **Asexual reproduction** leads to the rapid colonization of new areas, while **sexual reproduction** introduces genetic diversity, enhancing the fungi's ability to adapt to changing environmental conditions. Both modes of reproduction contribute to the ecological success of fungi as decomposers, pathogens, and symbionts in many ecosystems.

10

Bacteria as Plant Pathogens

Abstract

Bacteria, as single-celled prokaryotic organisms, include many species that act as plant pathogens causing significant crop losses worldwide. This chapter discusses the fundamental characteristics and morphology of phytopathogenic bacteria, highlighting their diverse shapes, cellular structures, and metabolic traits. The life cycle of bacterial pathogens involves infection, colonization, pathogenicity, and transmission through various vectors and environmental means. Classification is detailed based on morphology, Gram staining, pathogenicity, host range, transmission mode, and biochemical traits. Common bacterial diseases such as bacterial wilt, fire blight, bacterial spot, and crown gall are described along with their symptoms and management strategies. Effective control of bacterial diseases requires integrated approaches combining cultural, chemical, and biological methods to ensure sustainable crop health and productivity.

***Keywords**: Phytopathogenic bacteria, bacterial plant diseases, morphology, disease management, integrated disease management*

Introduction

Bacteria are single-celled prokaryotic organisms that can be both beneficial and harmful to plants. While many bacteria play essential roles in soil health and plant nutrition, certain pathogenic bacteria can cause severe diseases, leading to significant agricultural losses. Understanding the characteristics, life cycles, and management strategies for bacterial plant pathogens is crucial for effective disease control.

Characteristics and Morphology of Phytopathogenic Bacteria

General Characteristics

Phytopathogenic bacteria possess unique features that distinguish them from non-pathogenic strains:

- **Cell Structure:** Bacterial cells lack a true nucleus and membrane-bound organelles. Their genetic material is located in a single circular chromosome.

- **Cell Wall Composition:** Most phytopathogenic bacteria have a rigid cell wall composed of peptidoglycan, which provides structural support and protection. The composition of the cell wall varies, leading to classifications as Gram-positive or Gram-negative bacteria.
- **Metabolism:** Bacteria exhibit diverse metabolic capabilities. Some are autotrophic (able to produce their own food), while others are heterotrophic (requiring organic matter for nutrition).

Morphology

Shape of Plant Pathogenic Bacteria:

a) **Cocci (Spherical Bacteria):** Cocci-shaped plant pathogens are relatively less common but still significant in certain plant diseases. **Eg. *Xanthomonas* spp.,** responsible for bacterial blight in rice and bacterial spot in tomatoes and peppers. These bacteria cause water-soaked lesions and necrosis in plant tissues.

b) **Bacilli (Rod-shaped Bacteria):** Bacilli are more commonly found among plant pathogens and are responsible for various important diseases. **Eg. *Pseudomonas syringae***, known for causing bacterial speck on tomatoes, and *Pseudomonas savastanoi*, which causes olive knot disease. These bacteria produce toxins that interfere with plant defenses. Another example is *Erwinia amylovora*, the causative agent of fire blight in apple and pear trees, is a rod-shaped bacterium that causes wilting and blackening of blossoms and shoots.

c) **Spirilla (Spiral-shaped Bacteria):** Though less common in plant pathology, spiral-shaped bacteria can still affect certain plants. **Eg. *Spiroplasma citri*,** a spiral-shaped bacterium that causes citrus stubborn disease, leading to stunted growth, fruit deformities, and yellowing of leaves.

d) **Vibrio (Comma-shaped Bacteria):** While vibrio bacteria are more typically associated with aquatic environments, some plant pathogens resemble this shape. **Eg. *Phytoplasma*** (though not a true vibrio, it is pleomorphic and shares some structural similarities). *Phytoplasma* is responsible for diseases like aster yellows and can cause severe stunting and leaf deformation.

e) **Coccobacilli:** Intermediate in shape between cocci and bacilli, these bacteria also cause serious plant diseases. **Eg. *Ralstonia solanacearum*,** the agent behind bacterial wilt in solanaceous crops like tomatoes and potatoes. This bacterium causes wilting and plant death by colonizing the plant's vascular system.

Arrangement of Plant Pathogenic Bacteria

a) **Strepto (Chains of Bacteria):** Certain rod-shaped plant pathogenic bacteria can form chain-like arrangements. **Eg. *Streptomyces scabies*,** known for causing potato scab. These filamentous bacteria grow in soil and attack the tubers, leading to rough, scab-like lesions.

b) **Staphylo (Clusters of Bacteria):** Staphylococcus-like arrangements are uncommon in plant pathogens, but some bacteria might form irregular clusters under specific conditions. These arrangements may help them evade plant immune responses.

c) **Diplococci (Pairs of Bacteria):** While not common among plant pathogenic bacteria, a few could potentially exhibit paired arrangements. However, most plant pathogens tend to form longer chains or singular rods.

d) **Tetrads and Sarcina (Cube-like Packets):** These arrangements are more typically seen in environmental bacteria than in plant pathogens.

e) **Palisades:** Some bacilli arrange themselves in side-by-side formations. **Eg. *Corynebacterium michiganense*** (now known as *Clavibacter michiganensis*), responsible for bacterial canker in tomatoes. This bacterium can exhibit palisade-like arrangements and causes wilting, stem cankers, and fruit lesions.

Reproduction and Life Cycle of Bacterial Pathogens

Bacteria reproduce asexually through binary fission, a process where a single cell divides into two identical daughter cells. The life cycle of bacterial pathogens involves several stages:

1. **Infection:** Bacteria can enter plants through wounds, natural openings (stomata, lenticels), or directly through the root system.
2. **Colonization:** Once inside, bacteria colonize plant tissues and establish themselves, often forming biofilms that aid in their survival and pathogenicity.
3. **Pathogenicity:** Bacterial pathogens produce various virulence factors, such as toxins and enzymes, that contribute to their ability to cause disease. These factors can lead to cell death, wilting, and other symptoms.
4. **Transmission:** Bacterial diseases can spread through infected plant material, water, soil, and insect vectors. Effective control measures focus on breaking this transmission cycle.

Classification of Phytopathogenic Bacteria

1. **Classification Based on Morphology:** Bacterial morphology refers to the shape and physical structure of bacteria. It plays a crucial role in their identification.

   a) **Cocci (Spherical):** These bacteria are round in shape but are relatively rare as phytopathogens. **Eg. *Xanthomonas* spp.** (causes bacterial blight in rice).

   b) **Bacilli (Rod-shaped):** Rod-shaped bacteria are the most common form of plant pathogenic bacteria. **Eg. *Pseudomonas syringae*** (bacterial speck), *Erwinia amylovora* (fire blight), *Ralstonia solanacearum* (bacterial wilt).

   c) **Spirilla (Spiral-shaped):** These bacteria have a spiral shape and are typically motile. **Eg. *Spiroplasma citri*** (causes stubborn disease in citrus).

   d) **Coccobacilli:** Intermediate in shape between cocci and bacilli, these bacteria are slightly oval-shaped. **Eg. *Ralstonia solanacearum*** (causes bacterial wilt in a wide variety of plants).

2. **Classification Based on Gram Staining:** Gram staining differentiates bacteria into two groups based on the structure of their cell walls.

   a) **Gram-negative Bacteria:** These bacteria do not retain the crystal violet dye during Gram staining due to their thin peptidoglycan layer. They often have an outer membrane containing lipopolysaccharides. **Eg. *Xanthomonas campestris*** (black rot of crucifers), *Pseudomonas syringae* (bacterial speck), *Erwinia amylovora* (fire blight), *Ralstonia solanacearum* (bacterial wilt), *Agrobacterium tumefaciens* (crown gall).

   b) **Gram-positive Bacteria:** These bacteria retain the crystal violet dye, appearing purple after Gram staining due to their thick peptidoglycan layer. **Eg. *Clavibacter michiganensis*** (bacterial canker in tomatoes), *Streptomyces scabies* (potato scab), *Leifsonia xyli* (sugarcane ratoon stunting).

3. Classification Based on Pathogenicity: Phytopathogenic bacteria can be classified based on the nature of the disease they cause and the type of host plants they infect.

   a) **Vascular Pathogens:** These bacteria invade and colonize the plant's vascular system, typically the xylem or phloem. They interfere with water and nutrient transport, leading to systemic wilting and plant death. **Eg. *Ralstonia solanacearum*** (bacterial wilt), *Clavibacter* *michiganensis* (bacterial canker), *Xylella fastidiosa* (Pierce's disease in grapevines).

b) **Necrotrophic Pathogens:** These bacteria kill plant cells and feed on dead tissues, often causing lesions and necrosis. **Eg.** ***Pectobacterium carotovorum*** (soft rot of vegetables), *Dickeya spp.* (bacterial soft rot), *Erwinia amylovora* (fire blight).

c) **Biotrophic Pathogens:** These bacteria establish a long-term interaction with living plant cells, typically causing little or no immediate cell death. **Eg.** ***Xanthomonas oryzae*** *pv.* ***oryzae*** (bacterial blight of rice), *Agrobacterium tumefaciens* (crown gall disease), *Pseudomonas syringae* (various specks and blights).

4. **Classification Based on Host Range:** Plant pathogenic bacteria can also be classified based on the range of plants they infect.

a) **Host-specific Pathogens:** These bacteria infect only a specific host or a narrow range of hosts. **Eg.** ***Xanthomonas oryzae*** *pv. oryzae* (bacterial blight of rice), *Pseudomonas syringae pv. tomato* (bacterial speck in tomato), *Agrobacterium vitis* (crown gall in grapevines).

b) **Non-host-specific (Polyphagous) Pathogens:** These bacteria can infect a wide range of plant species from different families. **Eg.** ***Ralstonia solanacearum*** (infects over 200 plant species, including tomatoes, potatoes, and bananas), *Pectobacterium carotovorum* (causes soft rot in various vegetables), *Erwinia chrysanthemi* (soft rot and wilting in many plant species).

Classification Based on Mode of Transmission: Plant pathogenic bacteria can be classified based on how they spread from plant to plant.

a) **Soil-borne Pathogens:** These bacteria reside in the soil and infect plants through the root system. **Eg.** ***Ralstonia solanacearum*** (bacterial wilt), *Agrobacterium tumefaciens* (crown gall), *Clavibacter michiganensis* (bacterial canker).

b) **Water-borne Pathogens:** These bacteria spread through irrigation water, rain splash, or surface water. **Eg.** ***Xanthomonas campestris*** (black rot in crucifers), *Erwinia amylovora* (fire blight).

c) **Seed-borne Pathogens:** Some bacteria are transmitted through contaminated seeds, infecting plants during germination. **Eg.** ***Clavibacter michiganensis*** *subsp. sepedonicus* (ring rot of potato), *Pseudomonas syringae pv. syringae* (affects multiple crops like beans and wheat).

d) **Insect-vectored Pathogens:** Certain phytopathogenic bacteria rely on insects for transmission. **Eg.** ***Xylella fastidiosa*** (transmitted by sharpshooters in grapevines), *Spiroplasma citri* (transmitted by leafhoppers in citrus).

6. **Classification Based on Biochemical & Physiological Characteristics:** Phytopathogenic bacteria are also classified based on their biochemical activities, such as their metabolic needs, toxin production, and enzyme activity.
   a) **Aerobic Bacteria:** These bacteria require oxygen for survival and are generally involved in surface infections. **Eg. *Xanthomonas* spp.** (bacterial blight), *Pseudomonas syringae* (bacterial speck), *Erwinia spp.* (fire blight and soft rot).
   b) **Anaerobic Bacteria:** These bacteria thrive in oxygen-limited environments and are involved in diseases that affect underground parts like roots and tubers. **Eg. *Clostridium*** spp. (causes diseases in submerged plants or root crops).
   c) **Toxin-producing Bacteria:** Some plant pathogenic bacteria produce toxins that disrupt normal plant processes. **Eg. *Pseudomonas syringae*** *pv.* ***phaseolicola*** (produces phaseolotoxin, which causes halo blight in beans), *Xanthomonas spp.* (produces various toxins that lead to tissue necrosis).

Common Bacterial Diseases and Their Management Strategies

1. **Bacterial Wilt:** Caused by *Ralstonia solanacearum*, bacterial wilt affects crops like tomato, potato, and tobacco. The disease begins with wilting of the youngest leaves during the hottest part of the day, followed by recovery at night. As it progresses, wilting becomes permanent, and the leaves turn yellow. Browning of vascular tissues is commonly observed, and when stems are cut, a slimy, milky exudate may ooze out. In advanced stages, the entire plant collapses and dies. Management strategies include using resistant cultivars, practicing crop rotation with non-susceptible crops, improving soil drainage, and avoiding waterlogged conditions. Solarization of soil and sanitation of tools are also recommended.
2. **Fire Blight:** *Erwinia amylovora* causes fire blight, a serious disease affecting apples, pears, and other ornamental plants. Symptoms begin as water-soaked spots on blossoms and young shoots, which soon turn black and shriveled. As the disease spreads, entire branches become blackened and appear scorched, bending into a characteristic "shepherd's crook" shape. In severe cases, large sections of the tree turn black, giving a burnt appearance. Effective management includes pruning infected branches

below visible symptoms, applying copper-based bactericides during the growing season, avoiding excessive nitrogen fertilization, and planting fire blight-resistant varieties.

3. **Bacterial Spot:** Caused by *Xanthomonas* spp., bacterial spot affects tomatoes and peppers. The disease starts with small, dark, water-soaked spots on leaves, often surrounded by yellow halos. These lesions enlarge, turning brown or black. Affected leaves may yellow, dry up, and fall prematurely. Fruits develop rough, pitted lesions, leading to shriveling and deformation. Management strategies include using certified disease-free seeds, rotating crops to reduce pathogen build-up, applying copper-based fungicides, and avoiding overhead irrigation to minimize leaf wetness.
4. **Bacterial Blight:** *Pseudomonas syringae* causes bacterial blight, which affects crops like beans, cereals, and stone fruits. The disease begins with water-soaked lesions on leaves that turn brown or black with yellow margins. These lesions enlarge, causing curling and shriveling of leaves. Large necrotic areas may form, leading to leaf drop, stem cankers, and, in some cases, death of branches. Management includes using certified disease-free seeds, rotating crops, removing infected plant debris, applying copper-based bactericides early in the growing season, and avoiding planting in areas with poor air circulation.
5. **Soft Rot:** Caused by *Pectobacterium carotovorum*, soft rot primarily affects root vegetables like potatoes and carrots. The disease is characterized by soft, water-soaked spots that develop on affected plant tissues, especially in stems and tubers. As it advances, the tissue breaks down into a slimy, foul-smelling mass. In severe cases, the entire infected tissue disintegrates into a mushy, rotting mess. Management involves ensuring proper post-harvest storage conditions (cool and dry), avoiding mechanical damage during harvesting, using bactericides during the growing season, and planting in well-drained soil to avoid waterlogging.
6. **Crown Gall:** *Agrobacterium tumefaciens* causes crown gall, a disease that leads to the formation of small, rough, white galls near the soil line on stems or roots. These galls enlarge over time, becoming woody and disfigured. The disease stunts plant growth and affects nutrient uptake, leading to overall plant decline. Management includes avoiding wounding plants during cultivation, using disease-free planting material, treating infected soil with biological control agents like *Agrobacterium radiobacter*, and practicing crop rotation with non-susceptible plants.

7. **Bacterial Canker:** *Clavibacter michiganensis* subsp. *michiganensis* causes bacterial canker in tomatoes. The disease starts with wilting of individual leaflets, followed by the development of small, white lesions along the veins. As the infection advances, cankers form on stems and branches, followed by yellowing and necrosis of tissues. Fruits may shrivel and deform, leading to branch dieback. Management strategies include using resistant cultivars, avoiding overhead irrigation, practicing strict sanitation, disinfecting tools and equipment, and rotating with non-host crops.
8. **Angular Leaf Spot:** *Pseudomonas syringae* pv. *lachrymans* causes angular leaf spot in cucurbits like cucumbers, melons, and squash. Angular, water-soaked lesions appear on leaves, confined by leaf veins, giving the lesions their characteristic angular shape. As the disease progresses, the infected leaves become tattered and may fall off. Management strategies include rotating crops, removing infected plant debris, applying bactericides early in the growing season, avoiding excessive irrigation, and using resistant varieties.
9. **Bacterial Scab:** *Streptomyces scabies* causes bacterial scab, affecting root vegetables like potatoes, beets, and radishes. The disease leads to rough, corky lesions on the surface of tubers or roots, ranging from shallow to deep pits. These lesions reduce marketability and quality. Management strategies include avoiding planting in alkaline soils, using certified disease-free seed potatoes, rotating with non-host crops, and applying organic matter to improve soil health and reduce scab incidence.
10. **Bacterial Leaf Spot of Pomegranate:** *Xanthomonas axonopodis* pv. *punicae* causes bacterial leaf spot in pomegranate. Dark, water-soaked spots appear on leaves, turning into necrotic lesions. In advanced stages, the lesions spread to the fruits, causing unsightly spots and reducing fruit quality. Management involves using disease-free planting material, pruning affected areas to prevent spread, applying copper-based fungicides as a preventive measure, and avoiding excessive overhead irrigation.

Conclusion

Phytopathogenic bacteria are significant contributors to plant diseases, affecting agricultural productivity and ecosystem health. Understanding their characteristics, life cycles, and the diseases they cause is essential for developing effective management strategies. Integrated approaches that combine cultural, chemical, and biological methods are crucial for minimizing the impact of bacterial diseases on crops.

11

Other Plant Pathogens

Abstract

Chapter 11 explores a diverse group of lesser-known plant pathogens beyond the commonly studied fungi, bacteria, viruses, and viroids. It highlights the significance of Mollicutes (phytoplasmas and spiroplasmas), flagellate protozoa, fastidious vascular bacteria, parasitic algae, and parasitic higher plants in plant pathology. Each group is characterized by unique biological features and mechanisms of infection, often causing severe diseases that impact agricultural productivity and ecosystem health. The chapter discusses the biology, disease symptoms, transmission methods, and management strategies for these pathogens. Understanding these organisms is essential for developing integrated disease management approaches, especially given their potential to cause devastating plant diseases and their complex interactions within plant and insect hosts.

Keywords: *Mollicutes, Phytoplasmas, Spiroplasmas, Flagellate Protozoa, Fastidious Vascular Bacteria*

In addition to fungi, bacteria, viruses, and viroids, several other lesser-known groups of organisms also play significant roles in plant pathology. These include Mollicutes, protozoa, fastidious vascular bacteria, parasitic algae, and parasitic higher plants. Though these groups are not as commonly discussed, they can cause severe diseases in plants and should not be overlooked in integrated disease management strategies.

A. Mollicutes: Phytoplasmas and Spiroplasmas

Characteristics of Mollicutes

Mollicutes are a group of bacteria-like organisms that lack a cell wall, making them more flexible and pleomorphic (variable in shape). They are primarily obligate parasites, meaning they can only survive within a host. The two most significant genera within this group are phytoplasmas and spiroplasmas.

a) **Phytoplasmas:** These are wall-less, pleomorphic bacteria that reside in the phloem of infected plants, causing a range of systemic diseases. Phytoplasmas are transmitted primarily by phloem-feeding insect vectors, such as leafhoppers.

b) **Spiroplasmas:** These are spiral-shaped, wall-less bacteria that also reside in the phloem and are transmitted by insect vectors.

Phytoplasma-Induced Diseases

Phytoplasmas cause diseases that often lead to severe deformities in plants. Some of the major diseases caused by phytoplasmas include:

a) **Aster Yellows:** This disease affects a wide range of plants, including carrots, lettuce, and celery. Symptoms include yellowing of leaves, stunting, and phyllody (the abnormal development of floral parts into leafy structures).

b) **Coconut Lethal Yellowing:** A deadly disease affecting coconut palms, causing premature fruit drop, yellowing of leaves, and eventual death of the plant.

Management of Phytoplasma Diseases : Management strategies primarily include:

1. Vector control through insecticides or biological control agents.
2. Use of resistant plant varieties.
3. Removal of infected plants to reduce the source of inoculum.

B. Flagellate Protozoa

Flagellate Protozoa are a group of single-celled organisms that move using one or more whip-like structures called flagella. These protozoa are part of the larger phylum Mastigophora, characterized by their ability to propel themselves through liquid environments using their flagella. Flagellates can be free-living or parasitic, and they inhabit a wide range of environments, including soil, freshwater, and marine ecosystems.

Role in Plant Disease Transmission

Flagellate protozoa play a significant role in plant disease transmission, particularly in facilitating the spread of certain plant pathogens. While they are not commonly known as primary plant pathogens themselves, they can serve as vectors or indirect agents in disease cycles:

1. **Transmission of Plant Pathogenic Bacteria**: Some flagellate protozoa assist in spreading bacteria that cause plant diseases. By feeding on bacterial populations in the soil or plant rhizosphere, they can redistribute bacterial cells to different locations, facilitating the spread of pathogens like *Ralstonia* and *Xanthomonas*.
2. **Association with Root Diseases**: Flagellates can interact with pathogenic fungi or oomycetes in the rhizosphere, indirectly enhancing plant disease severity. By feeding on beneficial soil microbes, flagellate protozoa

may reduce microbial competition, allowing soilborne pathogens like *Phytophthora* and *Pythium* to infect plant roots more easily.

3. **Enhancing Soil Pathogen Activity**: In some cases, protozoa grazing on bacteria can disrupt the balance of microbial communities, promoting conditions favourable for plant pathogens. This can lead to an increase in the presence of disease-causing organisms in the soil, indirectly promoting root infections.
4. **Role in Nutrient Cycling**: By feeding on bacteria, flagellates contribute to nutrient cycling in the soil. Their activities can affect the availability of nutrients for plants, sometimes weakening plant defense mechanisms and making them more susceptible to infections by plant pathogens.

Examples of Protozoan-Transmitted Diseases

An important disease transmitted by protozoa is:

Peanut Clump Disease: This disease is caused by a soilborne virus, transmitted by protozoa, which results in stunted growth and clumping of peanut plants.

Management of Protozoan-Related Diseases

Managing protozoa-transmitted diseases can be challenging. Effective strategies often focus on:

1. Soil fumigation to reduce the protozoa population.
2. Crop rotation to prevent the buildup of the virus in the soil.
3. Use of resistant crop varieties where available.

C. Fastidious Vascular Bacteria (FVB)

Fastidious Vascular Bacteria (FVB) are a group of specialized bacteria that predominantly inhabit the vascular tissues of plants, including the xylem and phloem. These bacteria are known for their complex nutritional requirements and dependency on specific environmental conditions, making them challenging to culture and study in laboratory settings. They are often associated with significant plant diseases, causing symptoms such as wilting, yellowing, and stunted growth, which can lead to substantial agricultural losses.

Characteristics of Fastidious Vascular Bacteria (FVB)

1. **Slow Growth Rate**: FVB typically exhibit slow growth in culture, requiring extended incubation times for observable growth.
2. **Nutritional Dependency**: These bacteria have complex nutritional requirements, often needing specific nutrients or growth factors that are difficult to provide in standard culture media.

3. **Fastidiousness**: FVB are termed "fastidious" because they are difficult to culture outside their natural host environments, requiring living host tissues or specialized media for growth.
4. **Host-Specificity**: Many FVB are highly specific to their host plants, often causing disease in particular species or genera, and they may depend on host-specific factors for survival and virulence.
5. **Vascular Localization**: They primarily colonize vascular tissues, where they can move within the plant's transport systems, impacting the plant's ability to transport water and nutrients.
6. **Transmission Vectors**: FVB are often transmitted by specific insect vectors, such as leafhoppers, sharpshooters, or psyllids, which facilitate their spread from infected to healthy plants.
7. **Molecular Techniques for Identification**: Due to their fastidious nature, FVB are commonly identified using molecular techniques, such as PCR and DNA sequencing, rather than traditional culture methods.
8. **Pathogenicity**: Many FVB are known to cause serious diseases in plants, often leading to significant economic losses in agriculture.
9. **Biofilm Formation**: Some FVB can form biofilms in plant tissues, which can protect them from plant defenses and enhance their survival.

Major Diseases Caused by Fastidious Vascular Bacteria

1. **Pierce's Disease of Grapevine** (caused by *Xylella fastidiosa*): This disease blocks the xylem vessels, leading to water stress, leaf scorch, and plant death.
2. **Citrus Greening or Huanglongbing** (caused by *Candidatus Liberibacter spp.*): This is one of the most destructive citrus diseases, causing yellowing of leaves, stunted growth, poor fruit quality, and eventual death of the tree.

Management of Fastidious Vascular Bacteria

1. Vector control to reduce transmission.
2. Use of resistant rootstocks and planting materials.
3. Cultural practices like removal of infected plants to reduce sources of inoculum.

D. Parasitic Algae

Parasitic Algae are a specialized group of algae that derive their nutrients and energy from a host organism, often leading to the detriment of the host. Unlike free-living algae, which can photosynthesize and obtain energy from sunlight, parasitic algae rely on their hosts for survival. These organisms can infect various types of hosts, including aquatic plants, other algae, and even terrestrial plants.

Characteristics of Parasitic Algae

1. **Nutritional Dependency**: Parasitic algae obtain their nutrients and energy directly from their host organisms, making them dependent on the host for survival and growth.
2. **Lack of Photosynthesis**: Many parasitic algae either lack chlorophyll or have reduced photosynthetic capabilities compared to their free-living counterparts, leading them to rely entirely on their hosts for sustenance.
3. **Host Specificity**: Parasitic algae often show a degree of host specificity, meaning they can infect only certain types of organisms or specific species.
4. **Reproductive Strategies**: These algae can reproduce both sexually and asexually, often producing spores that can spread to new hosts. Their reproductive methods may depend on the host's life cycle.
5. **Invasive Nature**: Parasitic algae can rapidly colonize and invade host tissues, which may lead to the host's decline or death. This invasive ability can cause significant ecological and economic impacts, especially in aquatic ecosystems.
6. **Morphological Adaptations**: Parasitic algae often exhibit morphological adaptations, such as specialized structures for attachment to host tissues, enabling them to penetrate and extract nutrients effectively.
7. **Symbiotic Relationships**: In some cases, parasitic algae can form symbiotic relationships with their hosts, where both organisms may benefit or suffer from the interaction depending on environmental conditions.
8. **Diverse Taxonomy**: Parasitic algae belong to various algal groups, including red algae (Rhodophyta), green algae (Chlorophyta), and certain dinoflagellates.
9. **Pathogenicity**: Many parasitic algae can cause diseases in their hosts, leading to symptoms such as discoloration, necrosis, and reduced growth, which can ultimately impact the health of ecosystems or agricultural crops.

E. Parasitic Higher Plants

Parasitic Higher Plants are specialized flowering plants that obtain some or all of their nutrients and water from other living plants (hosts) rather than from soil. These plants have evolved various adaptations that allow them to attach to and penetrate the tissues of their hosts, drawing resources for their growth and development. Parasitic higher plants can be classified into two main categories: **hemiparasites**, which can photosynthesize and obtain some nutrients from the host, and **holoparasites**, which are fully dependent on their hosts and do not perform photosynthesis.

Characteristics of Parasitic Higher Plants

1. **Nutritional Dependency**: Parasitic higher plants derive nutrients, water, and sometimes carbohydrates directly from their host plants, often leading to nutrient depletion in the host.
2. **Attachment Structures**: These plants possess specialized structures called haustoria, which are modified roots that penetrate the host's tissues and connect to its vascular system, allowing for the transfer of nutrients.
3. **Diverse Taxonomy**: Parasitic higher plants belong to various taxonomic groups, including families such as **Orobanchaceae** (e.g., broomrape), **Loranthaceae** (e.g., mistletoe), and **Cuscuta** (dodder).
4. **Photosynthetic Ability**: Hemiparasitic plants can perform photosynthesis, allowing them to generate some energy independently, whereas holoparasitic plants lack chlorophyll and rely entirely on their hosts.
5. **Life Cycle and Reproduction**: Parasitic plants often have complex life cycles, including seed dispersal strategies that allow them to find and attach to suitable hosts. They can reproduce both sexually and asexually.
6. **Host Specificity**: Many parasitic higher plants show a degree of host specificity, meaning they preferentially attach to certain species or families of plants, which can impact the distribution and abundance of both the parasites and their hosts.
7. **Reduced Competition**: By relying on their hosts for resources, parasitic plants can often thrive in nutrient-poor environments where other plants may struggle to survive.
8. **Pathogenicity**: Parasitic higher plants can cause diseases in their hosts, leading to reduced growth, yield losses, and in severe cases, host death. Symptoms in hosts may include wilting, stunted growth, and yellowing of leaves.
9. **Ecological Impact**: Parasitic plants can significantly impact ecosystem dynamics, altering plant community structure, nutrient cycling, and competitive relationships between plants.

Major Parasitic Plants and Their Impact

Dodder (*Cuscuta* spp.)

- Holoparasitic vine that lacks chlorophyll and cannot perform photosynthesis.
- Wraps around host plants, establishing connections through haustoria that penetrate host tissues.
- Extracts water, carbohydrates, and nutrients, compromising the host's vascular system.
- Can severely weaken or kill its host, leading to stunted growth and diminished yield.
- Rapid spread creates a dense mat, reducing host access to sunlight and air.
- Particularly problematic in agricultural settings, infesting crops like legumes, cereals, and ornamentals, resulting in significant economic losses.

Mistletoe (*Viscum* spp. and *Phoradendron* spp.)

- Hemiparasitic plant that attaches to the branches of trees and shrubs.
- Possesses green leaves and can perform photosynthesis but relies on the host for water and nutrients.
- Extracts resources through haustoria that penetrate the host's bark, connecting to its vascular system.
- Reduces vigor of the host plant, leading to stunted growth and increased susceptibility to environmental stress.
- In forest ecosystems, mistletoe can provide food and habitat for wildlife while causing economic losses and reduced timber quality in agriculture.

Broomrape (*Orobanche* spp.)

- Holoparasitic plants that primarily infect the roots of crops like tomatoes, sunflowers, and tobacco.
- Lacks chlorophyll and relies entirely on the host for nutrients and water.
- Attaches to roots via haustoria, leading to stunted growth and reduced yields.
- Can cause significant economic losses in agricultural systems by decreasing crop health and productivity.
- Disrupts soil health and biodiversity by altering root interactions and nutrient cycling.

- Management is challenging due to its extensive seed bank and rapid spread, necessitating integrated strategies for control.

Management of Parasitic Plants

- Mechanical removal: Physically removing parasitic plants can reduce their spread, though it may be labor-intensive.
- Host resistance: Growing resistant varieties of crops can help minimize the impact of parasitic plants.
- Cultural practices: Techniques such as crop rotation, maintaining proper plant spacing, and avoiding planting in areas infested with parasitic plants can reduce the risk of infection.

12

Viruses and Viroids in Plant Pathology

Abstract

Chapter 12 focuses on viruses and viroids, two of the most significant and destructive pathogens in plant pathology. Viruses are submicroscopic infectious agents composed of nucleic acids enclosed in protein coats, which replicate exclusively within living host cells, causing systemic infections with symptoms such as mosaic patterns, leaf curling, chlorosis, necrosis, and stunting. Viroids, even simpler pathogens consisting solely of small circular RNA molecules without protein coats, disrupt host cellular processes and induce similar disease symptoms. The chapter explores their structure, replication mechanisms, transmission modes including insect vectors, mechanical means, seed, pollen, and grafting and highlights major viral diseases affecting important crops worldwide. Management strategies including cultural practices, resistant varieties, vector control, and sanitation are discussed to mitigate the impact of these pathogens on agricultural productivity.

***Keywords**: Plant viruses, Viroids, Plant disease symptoms, Transmission modes, Insect vectors, Resistant varieties, Integrated disease management*

Viruses and viroids are some of the most devastating pathogens affecting plants. They are sub-microscopic agents that can cause widespread damage to agricultural crops, leading to significant economic losses. Unlike fungi and bacteria, viruses and viroids are obligate parasites they rely entirely on a living host for replication and cannot survive independently in the environment. Their ability to rapidly spread and cause systemic infections makes them particularly challenging to manage.

A. Viruses: Structure and Characteristics

What is a Virus?

A virus is an infectious agent composed primarily of nucleic acid (either DNA or RNA) enclosed in a protein coat called a capsid. In some cases, viruses may also have an outer lipid membrane or envelope. Plant viruses are much smaller than bacteria and fungi, and they lack the cellular structures necessary for independent life. They can only replicate by hijacking the host cell's machinery.

Structure of Viruses

1. **Nucleic Acid:** This is the genetic material of the virus, which can be either RNA or DNA, and it carries the instructions for making new virus particles. Plant viruses are mostly RNA viruses, though some DNA viruses also exist.
2. **Capsid:** The protein shell that encloses and protects the nucleic acid. The capsid is made of protein subunits called capsomeres and can take various shapes, such as rod-like (rigid or flexible) or spherical.
3. **Envelope:** Some plant viruses may have an outer lipid envelope derived from the host cell's membrane, which provides additional protection.

Replication of Plant Viruses

Plant viruses enter host cells by mechanical damage or through insect vectors. Once inside, they disassemble and release their nucleic acid. This nucleic acid hijacks the host's cellular machinery to produce viral proteins and replicate the viral genome. The new virus particles are assembled within the host cell and then spread to adjacent cells via plasmodesmata (small channels between plant cells).

Symptoms of Viral Infections in Plants

Viral infections in plants often manifest through various distinctive symptoms that can significantly affect plant health and yield. Here are some common symptoms associated with viral infections in plants:

1. **Mosaic Patterns:** Mosaic patterns present as irregular patches of light and dark green, yellow, or white areas on leaves, creating a mottled appearance. These patterns occur due to uneven chlorophyll distribution caused by the virus's interference with chloroplast function. An example of this symptom can be seen in plants infected with Tobacco Mosaic Virus (TMV), which affects not only tobacco but also a wide range of other host plants like tomato and pepper.
2. **Leaf Curling or Distortion:** Leaf curling or distortion is characterized by abnormal twisting or curling of leaves, often accompanied by thickening or puckering of leaf tissue. This abnormal growth can significantly reduce the leaf area, ultimately impacting photosynthesis and overall plant health. For instance, the Tomato Curl Virus causes upward curling of tomato leaves, while Pepper Mottle Virus leads to similar curling symptoms in pepper plants.
3. **Yellowing (Chlorosis):** Yellowing, or chlorosis, occurs when leaves become discolored, often turning yellow due to damage to chloroplasts

and disruption of chlorophyll production. This symptom is detrimental to photosynthesis, leading to reduced plant vigor. Cucumber Mosaic Virus (CMV) is known to cause yellowing in cucumber plants, along with many other cucurbit crops.

4. **Stunting:** Stunting is marked by reduced growth and vigor of the plant, resulting in a smaller overall size compared to healthy counterparts. This symptom often leads to lower yields and can have serious implications for crop production. An example of stunting can be seen in plants infected with Bean Yellow Mosaic Virus, which leads to significantly reduced growth in bean plants.
5. **Necrosis:** Necrosis manifests as the death of plant tissue in localized areas, resulting in brown or black spots on leaves and stems. This tissue death can spread and severely weaken the plant. For example, Paprika Virus induces necrotic lesions on the leaves of infected pepper plants, which can further progress to significant tissue death.
6. **Leaf Spotting:** Leaf spotting involves the development of spots or lesions on leaves that vary in size and color, often leading to premature leaf drop. These spots may signify localized tissue damage that can disrupt the plant's ability to photosynthesize. Potato Virus Y causes dark brown or black spots on potato leaves, adversely affecting their health and leading to reduced tuber quality.
7. **Flower Abnormalities:** Flower abnormalities are characterized by distorted or malformed flowers, often resulting in reduced seed production and poor pollination. This symptom negatively impacts the reproductive success of the plant. Cucumber Mosaic Virus can cause flowers to become misshapen in infected cucumbers, leading to reduced fruit set and yield.
9. **Vein Clearing or Discoloration:** Vein clearing or discoloration is observed when the veins of leaves lose color, giving them a pale or yellow appearance while the interveinal areas may remain green. This symptom indicates a disruption in vascular function. Tomato Yellow Leaf Curl Virus results in noticeable vein clearing in infected tomato plants, severely impacting photosynthesis.
10. **Reduced Fruit Size and Quality:** Reduced fruit size and quality can occur in infected plants, leading to smaller, misshapen fruits that often have inferior quality. This symptom can severely impact marketability and consumption. An example of this is seen in plants infected with Tomato Mosaic Virus, which causes smaller, poorly developed tomatoes.

11. **Systemic Infection:** Systemic infection is characterized by symptoms appearing on leaves or stems distant from the initial site of infection, indicating that the virus has spread throughout the plant. This can lead to widespread damage and decline in health. Beet Yellow Virus is known to cause systemic symptoms in beet plants, resulting in overall yellowing and wilting.

B. Viroids: Structure and Characteristics

What is a Viroid?

A viroid is a much simpler pathogen than a virus, consisting solely of a short strand of circular RNA without a protein coat. Viroids are the smallest known agents of infectious diseases in plants. Unlike viruses, viroids do not code for any proteins; instead, they interfere with the host's cellular processes through their RNA structure.**Characteristics of Viroids**

1. Size: Viroids are much smaller than viruses, typically 200-400 nucleotides in length.
2. Lack of Capsid: Unlike viruses, viroids have no protective protein coat or capsid, which makes them unique among plant pathogens.
3. Replication: Viroids replicate within the host cell's nucleus or chloroplasts using the host's own enzymes. They use a rolling-circle mechanism to replicate their RNA and are able to spread through plasmodesmata to neighboring cells.

Symptoms of Viroid Infections

Viroid infections can result in symptoms similar to those of viral infections, such as:

1. Stunting: The plant shows reduced growth and vigor.
2. Leaf yellowing: Loss of green pigmentation due to cellular damage.
3. Leaf deformation: Abnormal development of leaves, often leading to a twisted appearance.

Modes of Transmission

a) **Insect Vectors:** One of the most common ways that plant viruses and viroids are transmitted is through insect vectors. Insects such as aphids, leafhoppers, whiteflies, and thrips feed on the sap of infected plants and carry viruses or viroids to healthy plants. There are two main types of vector transmission:

   Non-persistent transmission: The virus attaches to the mouthparts of the insect and is quickly transmitted when the insect feeds on a new plant.

Persistent transmission: The virus is ingested by the insect and circulates within its body before being transmitted to another plant. In some cases, the virus can replicate within the insect.

b) **Mechanical Transmission:** Mechanical damage to plants, such as wounds caused by pruning, cultivation, or even human handling, can provide entry points for viruses and viroids. The virus or viroid can spread from an infected plant to a healthy one via contaminated tools, hands, or equipment.

c) **Seed and Pollen Transmission:** Some plant viruses and viroids can be transmitted through infected seeds or pollen. Infected seeds give rise to infected plants, while infected pollen can spread the virus or viroid to new plants during pollination.

d) **Grafting** Grafting is a common horticultural technique used to propagate plants. If an infected plant is used as a rootstock or scion, the virus or viroid can spread to the grafted part, resulting in the transmission of the pathogen to the entire plant.

Major Plant Viral Diseases and Control Strategies

Tobacco Mosaic Virus (TMV)

Symptoms: Mosaic patterns appear as irregular patches of light and dark green on the leaves, giving a mottled appearance. Infected plants often exhibit stunted growth, with leaves curling upward and showing a leathery texture. Flowers may also develop deformations, and the overall yield can significantly decrease.

Management Strategies

- **Cultural Practices:** Implement crop rotation to break the virus's life cycle and avoid planting TMV-susceptible plants in infested fields. Regularly clean tools and equipment to prevent mechanical transmission of the virus.
- **Resistant Varieties:** Use TMV-resistant varieties to minimize the risk of infection.
- **Sanitation:** Remove and destroy infected plant debris after harvest to eliminate sources of the virus.
- **Vector Control:** Monitor and manage aphid populations using insecticides or beneficial insects like ladybugs.

Cucumber Mosaic Virus (CMV)

Symptoms: Infected plants display yellowing and curling of leaves, with characteristic mottled or mosaic patterns. Fruits may be distorted and have reduced size, while overall plant vigor declines, leading to stunted growth and poor yields.

Management Strategies

- **Resistant Varieties:** Plant CMV-resistant crops to reduce the likelihood of infection.
- **Insect Control:** Implement integrated pest management (IPM) strategies to control aphid populations that transmit CMV, using insecticidal soaps and neem oil as needed.
- **Cultural Practices:** Rotate crops and avoid planting susceptible species near infected plants.
- **Sanitation:** Clean equipment and tools to prevent the spread of the virus.

Tomato Yellow Leaf Curl Virus (TYLCV)

Symptoms: Leaves exhibit upward curling and a yellowing that starts from the leaf margins, creating a cupped appearance. Infected plants are often stunted, with fewer flowers and fruits, leading to significant yield loss.

Management Strategies

- **Resistant Varieties:** Use TYLCV-resistant tomato varieties to prevent disease establishment.
- **Vector Management:** Employ insecticides and introduce natural predators like parasitic wasps to control whiteflies, the main vectors of TYLCV.
- **Crop Management:** Avoid planting tomatoes in areas with known TYLCV outbreaks and practice crop rotation to reduce virus prevalence.
- **Sanitation:** Regularly inspect and remove infected plants to limit the spread of the virus.

Potato Virus Y (PVY)

Symptoms: Symptoms include mottling and yellowing of leaves, often accompanied by necrotic lesions. Infected plants may also show stunted growth and reduced tuber size, impacting overall yield significantly.

Management Strategies

- **Resistant Varieties:** Utilize PVY-resistant potato cultivars to reduce disease incidence.
- **Seed Certification:** Use certified disease-free seed potatoes to prevent introducing the virus into new fields.
- **Vector Control:** Implement IPM techniques to control aphids, including the use of reflective mulches to deter them.
- **Sanitation:** Clean tools and machinery to minimize the risk of mechanical transmission.

Barley Yellow Dwarf Virus (BYDV)

Symptoms: Infected plants exhibit stunted growth and yellowing of leaves, particularly the older leaves. The virus can cause reduced grain yield and poor quality, significantly affecting harvest.

Management Strategies

- **Resistant Varieties:** Plant BYDV-resistant cereal varieties to enhance resilience against the virus.
- **Insect Management:** Control aphid populations using insecticides and introducing beneficial insects like lacewings to reduce vector transmission.
- **Cultural Practices:** Rotate crops and avoid planting cereals in infested areas to disrupt the virus lifecycle.
- **Timely Planting:** Delay planting until after the peak aphid populations to reduce the risk of infection.

Beet Yellow Virus (BYV)

Symptoms: Leaves of infected plants develop a yellowing that may progress to necrosis. Symptoms also include stunted growth and reduced root size, leading to decreased yield and quality in crops like sugar beets.

Management Strategies

- **Resistant Varieties:** Use resistant varieties of sugar beets and related crops to limit infection risk.
- **Insect Control:** Manage aphid vectors through targeted insecticides and biological controls.
- **Crop Rotation:** Rotate crops to minimize the reservoir of the virus and avoid planting susceptible crops in previously infested areas.
- **Sanitation:** Regularly remove infected plants and debris from the field to reduce sources of the virus.

Pepper Mottle Virus (PMV)

Symptoms: Infected pepper plants show mottled and yellowing leaves, with significant leaf curling and distortion. This leads to reduced photosynthetic capacity and poor fruit development, impacting yield.

Management Strategies

- **Resistant Varieties:** Grow PMV-resistant pepper varieties to mitigate the impact of the virus.
- **Vector Management:** Implement controls for aphids and whiteflies through insecticides and promoting beneficial insects in the field.
- **Sanitation:** Remove and destroy infected plants to prevent further spread of the virus.
- **Cultural Practices:** Rotate crops and maintain a clean growing environment to minimize infection risk.

Soybean Mosaic Virus (SMV)

Symptoms: Symptoms include mottling, yellowing, and leaf distortion, which can lead to stunted plants and reduced seed yield. In severe cases, symptoms may also affect the quality of the harvested soybeans.

Management Strategies

- **Resistant Varieties:** Plant SMV-resistant soybean cultivars to enhance resistance against the virus.
- **Cultural Practices:** Rotate crops and avoid planting soybeans near sources of infection.
- **Vector Control:** Manage aphid populations using integrated pest management techniques, including natural predators.
- **Sanitation:** Remove infected plants and debris promptly to limit virus spread.

Rice Tungro Virus (RTV)

Symptoms: Infected rice plants exhibit yellowing and stunting, with the leaves curling and turning upward. Infected plants typically produce fewer tillers and reduced grain yield.

Management Strategies

- **Resistant Varieties:** Grow RTV-resistant rice varieties to reduce the risk of disease establishment.
- **Vector Control:** Manage planthopper populations using targeted insecticides and biological controls.

- **Cultural Practices:** Avoid planting rice in infected fields and practice crop rotation to minimize infection spread.
- **Field Management:** Schedule planting to avoid peak vector populations, which can significantly decrease the likelihood of infection.

Zucchini Yellow Mosaic Virus (ZYMV)

Symptoms: Leaves develop mottling, yellowing, and stunted growth. Infected fruits often show deformation and reduced size, leading to poor marketability and yield.

Management Strategies

- **Resistant Varieties:** Use ZYMV-resistant zucchini and squash varieties to prevent infection.
- **Vector Management:** Control aphid populations using IPM strategies and insecticidal soaps to limit virus transmission.
- **Cultural Practices:** Rotate crops and implement good sanitation practices to minimize the risk of infection.
- **Sanitation:** Regularly inspect and remove infected plants and debris from the growing area to reduce the spread of the virus.

13

Principles of Plant Disease Management

Abstract

Plant disease management is an essential aspect of sustainable agriculture aimed at minimizing crop losses caused by various pathogens, including fungi, bacteria, viruses, nematodes, and abiotic factors. This chapter outlines the fundamental principles underlying effective disease management, anchored on the Disease Triangle concept involving the host, pathogen, and environment. Strategies such as avoidance, exclusion, eradication, protection, utilization of resistant varieties, and therapeutic interventions are discussed in detail. Each approach highlights practical methods, including crop rotation, certified seed use, sanitation, chemical and biological controls, and genetic resistance. Together, these integrated practices enable effective disease control, enhancing crop productivity while maintaining environmental and economic sustainability.

Keywords: *Plant Disease Management, Principles, Resistant Varieties, Therapeutic Strategies, Integrated Pest Management (IPM)*

Managing plant diseases is a crucial part of agriculture and horticulture. Plant disease management focuses on reducing the impact of diseases caused by various pathogens, including fungi, bacteria, viruses, nematodes, and other biotic or abiotic agents. The primary goal is to protect crops from severe damage while ensuring environmental sustainability and economic viability.

Basic Concepts of Plant Disease Management

Plant disease management is built upon understanding the Disease Triangle, which consists of three components:

Host: A susceptible plant that can be affected by the pathogen.

Pathogen: A disease-causing agent like fungi, bacteria, or viruses.

Environment: The surrounding conditions, such as temperature, moisture, and wind, that influence disease development.

For a disease to develop, all three components must interact favourably. By manipulating any of these elements, disease outbreaks can be minimized.

Principles of Plant Disease Management

Avoidance

Avoidance strategies focus on preventing disease occurrence by altering the environment or planting practices.

a) **Site Selection**: Select fields with no previous history of specific pathogens. For instance, avoid planting potatoes in fields previously infected with *Phytophthora infestans*, as the pathogen can survive in the soil and debris.

b) **Optimal Planting Dates**: Adjust planting dates to escape disease cycles. For example, planting wheat before peak conditions for *Puccinia triticina* (causing leaf rust) can reduce the likelihood of infection.

c) **Crop Rotation**: Implementing a crop rotation system with non-host crops disrupts the life cycles of pathogens. Rotating soybeans with corn can mitigate root-knot nematode (*Meloidogyne spp.*) populations.

d) **Soil Management**: Soil solarization (covering the soil with clear plastic to trap solar energy) can reduce pathogen loads, including *Verticillium dahliae* (causing wilt diseases) by raising soil temperatures.

Exclusion

Exclusion strategies prevent the introduction and spread of pathogens into disease-free areas.

a) **Use of Certified Seed**: Employing certified disease-free seed minimizes the introduction of seed-borne pathogens such as *Cucumber Mosaic Virus* or *Fusarium oxysporum*.

b) **Sanitation Protocols**: Regular cleaning of tools and equipment, such as disinfecting pruning shears with 70% ethanol, prevents mechanical transmission of pathogens like *Xanthomonas campestris* (causing bacterial blight).

c) **Physical Barriers**: Employing physical barriers like row covers can protect against vectors of viral pathogens such as *Aphids*, which transmit *Barley Yellow Dwarf Virus*.

Eradication

Eradication focuses on eliminating existing sources of pathogens.

a) **Rogueing**: The practice of identifying and removing infected plants, such as those showing symptoms of *Tomato Spotted Wilt Virus*, is crucial in preventing further spread within the crop.

b) **Soil Disinfestation**: Methods such as steam sterilization of soil can eradicate soil-borne pathogens like *Rhizoctonia solani*, effectively reducing their viability.

c) **Destruction of Infected Debris**: Burning or deep burying infected plant debris (e.g., from *Sclerotinia sclerotiorum* causing white mold) reduces the pathogen's potential to survive and infect future crops.

Protection

Protection strategies involve chemical and cultural practices to shield plants from infections.

a) **Chemical Fungicides and Bactericides**: Fungicides like azoxystrobin can protect against fungal pathogens such as *Botrytis cinerea* (grey mold) by inhibiting spore germination. Similarly, using copper hydroxide can manage bacterial infections like *Bacterial Spot* in peppers.

b) **Cultural Practices**: Practices such as maintaining proper plant spacing improve air circulation and reduce humidity, limiting the spread of foliar diseases like *Powdery Mildew* caused by *Erysiphe spp.*

c) **Monitoring and Integrated Pest Management (IPM)**: Regular monitoring for signs of disease and employing IPM strategies (e.g., introducing beneficial insects like ladybugs to control aphid populations that transmit viruses) can significantly reduce pathogen pressure.

Resistant Varieties

Utilizing resistant plant varieties is a key component of integrated disease management.

a) **Genetic Resistance**: Employing crop varieties that exhibit specific resistance genes, such as the *R* genes in wheat varieties that confer resistance to *Puccinia graminis* (stem rust), is essential.

b) **Marker-Assisted Selection**: Breeding programs using marker-assisted selection can enhance resistance traits in crops. For example, incorporating resistance to *Tomato Yellow Leaf Curl Virus* (TYLCV) in tomatoes through genetic markers has proven effective.

c) **Education and Training**: Educating farmers on the selection and benefits of resistant varieties can enhance adoption rates. For instance, promoting drought-resistant maize varieties that are also resistant to *Maize Lethal Necrosis Virus* can mitigate both abiotic and biotic stresses.

Therapy

Therapeutic strategies are employed to manage symptoms in infected plants.

a) **Use of Plant Growth Regulators**: Applying substances like jasmonic acid can enhance systemic acquired resistance (SAR) in plants, making them less susceptible to pathogens like *Pseudomonas syringae*.

b) **Biocontrol Agents**: Utilizing biocontrol agents, such as *Bacillus subtilis* or *Trichoderma harzianum*, can suppress pathogens like *Fusarium* and enhance plant resilience by outcompeting the pathogen or inducing plant defense responses.

c) **Antimicrobial Treatments**: The application of natural antimicrobials, such as chitosan or essential oils, can reduce disease severity in crops affected by pathogens like *Fusarium wilt.*

By employing these specific strategies, farmers and plant pathologists can effectively manage a wide range of plant pathogens, contributing to enhanced crop productivity and sustainable agricultural practices.

14

Disease Management Using Chemicals

Abstract

Chemical control remains a cornerstone in managing plant diseases by utilizing synthetic and natural substances to inhibit or eradicate pathogens. This chapter provides a comprehensive overview of chemical disease management, detailing the types of chemicals such as fungicides, bactericides, nematicides, and fumigants. It explains the distinction between contact and systemic pesticides, emphasizing their modes of action and practical applications. Understanding these modes of action is crucial for effective and sustainable use, helping to prevent pathogen resistance and environmental harm. Various application methods, benefits, risks, and limitations of chemical control are discussed, alongside best practices for safe and efficient use. Finally, the chapter explores future trends aimed at reducing reliance on synthetic chemicals through the development of safer pesticides and integration with other disease management strategies.

Keywords: *Chemical control, Contact Fungicides, Systemic Fungicides, Mode of action*

Chemical control of plant diseases involves the use of synthetic or naturally occurring substances to kill or inhibit the growth of disease-causing pathogens. Though highly effective and widely used, chemical control must be applied judiciously due to concerns about environmental impact, human health, and the development of pathogen resistance. This chapter explores the types, modes of action, application methods, and challenges associated with chemical control in plant disease management.

Overview of Chemical Control

Chemical control is primarily achieved through the application of pesticides. The term encompasses fungicides (for fungi), bactericides (for bacteria), nematicides (for nematodes), and other chemicals targeting specific pathogens. Chemical control serves as an immediate solution, offering quick action against disease outbreaks.

Types of Chemicals Used in Disease Management

1. **Fungicides:** Used to control fungal pathogens, fungicides can either prevent infection (protectant fungicides) or cure infections after they

occur (systemic fungicides). Examples include copper-based fungicides like copper oxychloride and sulfur-based compounds.

2. **Bactericides:** These are designed to control bacterial pathogens. Common examples include streptomycin and copper compounds, which act against bacterial blights and spots. However, bactericides are less widely used than fungicides due to limited availability of effective products.
3. **Nematicides:** These chemicals target nematodes, which cause root diseases in crops. Carbofuran & oxamyl are commonly used nematicides.
4. **Fumigants:** These are volatile chemicals applied to soil to kill a wide range of soil-borne pathogens, including fungi, bacteria, and nematodes. Examples include methyl bromide and chloropicrin.

Contact vs. Systemic Chemicals

1. **Contact Pesticides:** These chemicals remain on the surface of plants and must come into direct contact with pathogens to be effective. They are most effective as preventative measures.
2. **Systemic Pesticides:** These are absorbed by the plant and move through its tissues, offering protection from within. Systemic fungicides, such as triazoles, are curative, meaning they can halt an infection after it has started.

Difference between Contact Pesticides & Systemic Pesticides

Characteristic	Contact Pesticides	Systemic Pesticides
Mode of Action	Remains on the plant surface; direct contact with pathogen required.	Absorbed by the plant and circulates within plant tissues.
Effectiveness	Preventative; effective before pathogen contact or infection.	Curative; can halt infection after it has started.
Persistence	Degrades quickly due to environmental factors; requires frequent reapplication.	Longer-lasting; stays in plant tissues, reducing the need for reapplication.
Selectivity	Broad activity; may affect non-target organisms.	More selective; targets specific pathogens but may develop resistance over time.
Application Timing	Applied before infection for prevention.	Applied during or after infection for cure.
Environmental Impact	Higher risk due to frequent reapplication and broader activity.	Lower impact due to selective action but could lead to resistance.
Examples	Captan (fungicide), Mancozeb (fungicide), Copper sulfate.	Triazoles (fungicide), Imidacloprid (insecticide), Phosphites.
Practical Use	Used to prevent diseases like leaf spots, downy mildew.	Used to cure diseases like powdery mildew, root rot.

Modes of Action

Understanding how chemicals work at the biological level is essential for their effective use. The mode of action describes how a chemical affects the target pathogen.

1. **Inhibition of Cell Wall Synthesis:** This mode of action targets the synthesis of key structural components like β-glucans or chitin, essential for cell wall integrity in fungi. When cell wall synthesis is inhibited, fungal cells lose their structural support, leading to cell lysis and death. Examples: Pyriofenone: Inhibits fungal cell wall formation, effective against powdery mildew in grapes & tomatoes. Isopyrazam: Effective against leaf blotch and rust in cereals. Echinoctandins (e.g., Caspofungin): Target β-glucan synthesis, commonly used in human medicine but relevant in fungal control.
2. **Disruption of Cellular Membranes:** These fungicides interfere with the integrity of fungal or bacterial cell membranes, causing leakage of essential cellular contents like ions and proteins, leading to pathogen death due to loss of metabolic control. Examples: Fluopyram: A succinate dehydrogenase inhibitor (SDHI), disrupts cell membranes and energy production, effective against *Botrytis* and *Erysiphe*. Fluxapyroxad: Disrupts membranes by inhibiting succinate dehydrogenase, used in cereals. Pydiflumetofen: SDHI effective against powdery mildew and rust fungi.
3. **Inhibition of Protein Synthesis:** Protein synthesis is critical for pathogen growth and survival. Fungicides and bactericides that inhibit protein synthesis stop the pathogen from producing essential proteins, halting their growth. Examples: Kasugamycin: Inhibits bacterial protein synthesis, effective against *Xanthomonas oryzae* causing bacterial leaf streak in rice. Oxytetracycline: Blocks protein synthesis, effective against fire blight in apples and pears. Streptomycin: Used to control bacterial diseases like bacterial spots and fire blight.
4. **Inhibition of Respiration:** This mode of action blocks the pathogen's ability to produce energy by targeting the electron transport chain in mitochondria. Without energy, the pathogen cannot carry out essential functions, leading to death.

   Examples: Bixafen: A succinate dehydrogenase inhibitor (SDHI), effective against rust and leaf spot in cereals. Fluindapyr: Inhibits respiration in fungi like frogeye leaf spot and white mold in soybeans. Strobilurins (e.g., Azoxystrobin, Pyraclostrobin): Inhibits mitochondrial respiration, effective against powdery mildew, rusts, and downy mildew.

Fenpicoxamid: Novel fungicide used in wheat to control diseases like septoria.

5. **Inhibition of Nucleic Acid Synthesis:** These chemicals inhibit the synthesis of DNA and RNA, essential for the replication and transcription processes. Without the ability to produce nucleic acids, pathogens cannot reproduce or maintain normal cellular function.

   Examples: Flonicamid: Inhibits nucleic acid synthesis in sap-feeding insects, with potential applications in controlling bacterial and fungal pathogens. Cyprodinil: A fungicide disrupting RNA synthesis, effective against pathogens like *Botrytis* in grapes and strawberries.

6. **Inhibition of Lipid Biosynthesis:** Lipids are key components of cell membranes. Inhibiting lipid biosynthesis leads to defective cell membranes and malfunctions in signalling, resulting in cell death.

   Examples: Fenhexamid: Inhibits ergosterol biosynthesis, effective against grey mold (*Botrytis*) in grapes. Prothioconazole: Targets lipid biosynthesis in fungi like *Fusarium* and *Sclerotinia.*

7. **Inhibition of Amino Acid Synthesis:** Amino acids are the building blocks of proteins. Inhibiting amino acid synthesis disrupts protein production, leading to pathogen growth inhibition and death.

   Examples: Isoniazid: Inhibits the synthesis of mycolic acid, important in bacterial cell walls.

8. **Inhibition of Signal Transduction Pathways:** Signal transduction pathways regulate biological processes. Inhibiting these pathways interferes with the pathogen's ability to adapt and survive, leading to death.

   Examples: Boscalid: Inhibits signal transduction pathways related to fungal growth, used in vegetables, grapes, and berries. **Cyflufenamid**: Disrupts signal transduction, effective against powdery mildew.

9. **Multi-Site Action:** Some fungicides and bactericides act on multiple sites within a pathogen, attacking various metabolic pathways and cellular structures, reducing resistance development.

   Examples: Fluazinam: A broad-spectrum fungicide with multi-site activity, used against late blight in potatoes and downy mildew in grapes. **Chlorothalonil**: A multi-site fungicide preventing spore germination and disrupting fungal enzymes, though being phased out in some regions.

Significance of Understanding Mode of Action in Plant Disease Management

1. **Prevention of Resistance**: Knowledge of different modes of action enables the development of rotation programs to prevent resistance in pathogens.
2. **Target-Specific Control**: By knowing specific MOAs, more precise targeting of pathogens is possible, minimizing off-target effects and environmental damage.
3. **Integrated Disease Management (IDM)**: Understanding MOA is crucial in IDM frameworks for selecting the most appropriate chemicals and integrating them with cultural and biological control methods.
4. **Environmental and Regulatory Compliance**: Newer molecules with specific MOAs are often designed to have a lower environmental impact, crucial for meeting regulatory standards.
5. **Sustainable Plant Health**: Proper MOA usage extends the lifespan of fungicides and bactericides in managing plant health across multiple crops.

Application Methods

The effectiveness of chemical control depends on proper application. Chemicals must reach the target pathogen at the right time and in the right concentration to be effective.

1. **Foliar Sprays:** Foliar sprays are the most common method of applying fungicides and bactericides. Chemicals are sprayed directly onto the leaves, stems, or flowers of plants, where they act against pathogens. Proper coverage of the plant is critical to ensure that the pesticide comes into contact with the pathogen.
2. **Soil Applications:** Some pesticides, particularly nematicides and fumigants, are applied directly to the soil to target soil-borne pathogens. Soil fumigation is often used before planting to eliminate pathogens in the root zone. These chemicals are injected into the soil or spread and incorporated using irrigation systems.
3. **Seed Treatments:** Seed treatments involve coating seeds with fungicides or bactericides before planting. This protects the young plant from seed-borne pathogens or soil-borne pathogens during the early stages of growth. Fungicides like thiram and captan are commonly used for seed treatment.
4. **Post-Harvest Treatments:** Post-harvest diseases can result in significant losses. To prevent this, fruits and vegetables are often treated with

fungicides after harvest. Chemicals like imazalil and thiabendazole are used to control storage diseases like fruit rots caused by fungi.

Benefits of Chemical Control

1. **Rapid Disease Suppression:** Chemical control provides immediate results, often halting disease progression within hours or days. This is particularly important during epidemics or sudden outbreaks when other control methods may not work as quickly.
2. **Wide Range of Target Pathogens:** Many chemical pesticides have broad-spectrum activity, meaning they can control a variety of pathogens. This makes them useful in complex cropping systems where multiple diseases may be present.
3. **Ease of Application:** Modern pesticides are designed for ease of use. With proper equipment and training, they can be applied efficiently over large areas, making them ideal for large-scale farming operations.

Risks and Limitations

1. **Environmental Impact:** Overuse or misuse of chemical pesticides can lead to significant environmental damage. Runoff from fields can contaminate water sources, affecting aquatic life and entering the food chain. Certain chemicals also pose risks to non-target organisms, including beneficial insects, birds, and mammals.
2. **Development of Resistance:** Pathogens can develop resistance to chemical pesticides through repeated exposure. Resistance occurs when a small population of the pathogen survives the treatment and passes on resistant traits to future generations. This reduces the efficacy of the chemical and necessitates the use of alternative products or higher doses, further exacerbating environmental issues. An example is the widespread resistance of powdery mildew fungi to strobilurin fungicides.
3. **Human Health Concerns:** Many chemicals used in agriculture are toxic to humans, especially if proper safety precautions are not followed. Long-term exposure to pesticides can lead to chronic health issues such as respiratory problems, skin disorders, or even cancer in extreme cases. Strict regulations govern the use of certain chemicals to mitigate these risks, but care must always be taken during application.
4. **Residue Concerns:** The presence of pesticide residues in food products is a major concern for consumers. Excessive residues can occur if chemicals are applied too close to harvest or at improper doses. Maximum residue limits (MRLs) are set by regulatory bodies to ensure that food products remain safe for consumption.

Best Practices for Chemical Control

To maximize the benefits of chemical control while minimizing the risks, the following best practices should be adopted:

1. **Rotation of Chemicals:** Alternating between different classes of chemicals with different modes of action helps prevent the development of resistance in pathogens.
2. **Targeted Application:** Applying pesticides only when necessary (based on disease scouting or forecasting) reduces environmental impact and costs.
3. **Proper Dosing:** Following the recommended doses and timing of application is crucial to avoid the overuse of chemicals.
4. **Use of Protective Equipment:** Farmers and applicators should wear appropriate protective gear, such as gloves, masks, and goggles, to minimize exposure to harmful chemicals.
5. **Compliance with Regulations:** Always follow local and international regulations concerning pesticide use to ensure safety for the environment, farm workers, and consumers.

Future of Chemical Control

The future of chemical control will likely focus on reducing dependency on synthetic chemicals while maintaining effective disease management. This can be achieved through:

1. **Development of Safer, Biodegradable Pesticides:** Advances in chemical engineering are leading to the creation of pesticides that break down more quickly in the environment, reducing their long-term impact.
2. **Increased Use of Biological Pesticides:** Biological pesticides derived from natural organisms are seen as a more sustainable alternative to synthetic chemicals. These include products based on microbes, plant extracts, or naturally occurring minerals.
3. **Integration with Other Methods:** Chemical control will continue to play a role in Integrated Disease Management (IDM), where it is used in combination with biological, cultural, and genetic resistance methods for a more balanced and sustainable approach.

15

Host Resistance in Disease Management

Abstract

Host resistance is a vital and sustainable approach to managing plant diseases by harnessing the plant's inherent ability to defend itself against pathogens. This chapter explores the concept of host resistance, highlighting its significance in reducing chemical dependency and providing long-term protection against diverse diseases. It details the two main types of resistance—vertical (race-specific) and horizontal (general) resistance—and their genetic and practical implications. The chapter also discusses breeding strategies, including conventional methods, molecular breeding, and genetic engineering, used to develop disease-resistant varieties. Challenges such as pathogen evolution and balancing resistance with other agronomic traits are examined. Finally, the chapter looks ahead at innovative tools like gene pyramiding, genomic selection, and gene editing, which promise to enhance the durability and spectrum of host resistance in crop protection.

***Keywords**: Host resistance, Disease management, Disease-resistant varieties, Pathoge evolution, Sustainable agriculture*

Host resistance refers to the inherent ability of plants to defend themselves against disease-causing pathogens. Instead of relying on external inputs such as chemicals or biological controls, plants themselves possess mechanisms that enable them to resist or tolerate infections. Developing and utilizing resistant crop varieties has become a cornerstone of sustainable disease management, offering long-term protection with fewer inputs.

This chapter delves into the concept of host resistance, its importance in disease management, the types of resistance, and the strategies used to breed disease-resistant varieties.

Concept of Host Resistance

Host resistance is the natural or acquired ability of a plant to either completely prevent pathogen infection or minimize the impact of infection. This trait has evolved in plants as a defense mechanism against various pathogens such as fungi, bacteria, viruses, nematodes, and other parasites. While chemical and cultural controls offer external disease management solutions, host resistance operates from within the plant.

Key aspects of host resistance include

1. Prevention of Pathogen Entry: Resistant plants may prevent pathogens from entering their tissues by forming physical or biochemical barriers.
2. Limitation of Pathogen Growth: Once pathogens have penetrated, the plant's immune system may slow or stop their spread, minimizing damage.
3. Tolerance of Infection: Some plants do not stop the pathogen from infecting them but tolerate the presence of the pathogen without suffering significant damage.

Importance of Host Resistance in Disease Management

1. **Long-Term Protection:** Unlike chemical treatments that need repeated application, disease-resistant varieties offer long-term, stable protection against specific pathogens. Once developed, resistant varieties can protect crops throughout their life cycle, providing a reliable defense.
2. **Reduced Dependence on Chemicals:** By planting resistant varieties, farmers can significantly reduce or eliminate the need for chemical fungicides, bactericides, or nematicides. This reduces production costs, environmental impact, and risks to human health.
3. **Environmentally Sustainable:** Resistant plants offer a more environmentally friendly solution to disease management. The widespread adoption of resistant varieties can reduce the need for synthetic chemicals, preserving soil and water quality, as well as biodiversity.
4. **Protection Against Multiple Diseases:** Certain plant varieties can be bred to resist multiple pathogens, providing broad-spectrum protection. For instance, some wheat varieties are resistant to various fungal diseases like rusts and powdery mildew.

Types of Host Resistance

Host resistance can be broadly classified into two categories based on its effectiveness and duration: vertical resistance and horizontal resistance. Both types have different applications and advantages.

A. Vertical Resistance (Race-Specific Resistance)

Vertical resistance, also known as race-specific resistance, is effective against specific strains or races of a pathogen. It is usually governed by single or a few genes that allow the plant to resist specific pathogens through a mechanism known as the gene-for-gene interaction.

Key Features of Vertical Resistance

1. Highly Specific: Works against specific strains (races) of the pathogen.
2. Controlled by Few Genes: Involves one or a few resistance (R) genes.
3. Strong but Short-Term: It can be highly effective in controlling disease, but pathogens may evolve to overcome the resistance, rendering it less effective over time.
4. Easy to Breed: Vertical resistance is relatively easy to identify and introduce into new varieties through traditional or modern breeding techniques.

Example: The Pi-ta gene in rice provides race-specific resistance against the *Magnaporthe oryzae* fungus, which causes rice blast disease. However, new pathogen races can emerge that overcome this gene's defense.

B. Horizontal Resistance (General Resistance)

Horizontal resistance, also called general or polygenic resistance, offers broad-spectrum defense against a wide range of pathogen strains. Unlike vertical resistance, horizontal resistance is governed by multiple genes and provides partial but durable resistance.

Key Features of Horizontal Resistance

1. Broad-Spectrum: Works against multiple races or strains of a pathogen.
2. Controlled by Many Genes: Involves multiple genes that each contribute to partial resistance.
3. Durable but Moderate: Offers long-term resistance that pathogens find difficult to overcome, though the level of protection may not be complete.
4. More Complex to Breed: Horizontal resistance is more challenging to breed because it involves multiple genetic factors and is usually less easily identified.

Example: In potatoes, horizontal resistance to late blight (caused by *Phytophthora infestans*) has been developed, offering long-term protection across various pathogen strains.

Breeding for Disease-Resistant Varieties

Developing disease-resistant varieties involves identifying resistance traits in plant populations and integrating them into commercially viable crop varieties. This can be achieved through conventional breeding, molecular breeding, or genetic engineering.

Conventional Breeding

Conventional breeding relies on crossing resistant varieties or wild relatives with susceptible varieties to introduce resistance genes. This method is time-consuming but has been widely successful in crops such as wheat, rice, and maize.

Steps in Conventional Breeding

1. Selection of Resistant Parent: Identify a resistant plant variety or wild relative.
2. Crossing: Cross the resistant plant with a susceptible, high-yielding variety.
3. Screening: Select offspring that exhibit resistance traits.
4. Stabilization: Breed the selected plants over several generations to stabilize the resistance trait.

Molecular Breeding and Marker-Assisted Selection (MAS)

Molecular breeding involves using genetic markers to identify and select resistant plants at the seedling stage, accelerating the breeding process. Marker-assisted selection (MAS) enables breeders to quickly and accurately select plants carrying specific resistance genes without waiting for the plant to reach maturity or show disease symptoms.

Genetic Engineering

Genetic engineering allows scientists to directly introduce resistance genes from one organism into another, even if they are unrelated. This technique has enabled the development of transgenic crops with enhanced resistance to diseases.

Example: Bt Cotton: Genetically modified cotton plants containing genes from *Bacillus thuringiensis* (Bt) are resistant to bollworms, reducing the need for insecticides.

Challenges in Host Resistance Development

While host resistance is a powerful tool in disease management, it comes with challenges that need to be addressed.

Pathogen Evolution

Pathogens are capable of evolving to overcome host resistance. In the case of vertical resistance, pathogens may mutate or recombine, creating new races that can bypass the plant's defense mechanisms.

Trade-offs with Other Traits

Breeding for resistance can sometimes lead to trade-offs with other desirable traits like yield, quality, or drought tolerance. Finding a balance between disease resistance and these other traits is crucial for successful crop development.

Genetic Diversity

A reliance on a narrow range of resistant varieties can reduce genetic diversity in crops, making them more vulnerable to new pathogen strains. Maintaining genetic diversity in breeding programs is essential for the long-term success of disease resistance strategies.

Case Studies of Host Resistance in Disease Management

Wheat Rust Resistance

Wheat rust, caused by various species of *Puccinia*, is a significant problem for global wheat production. Researchers have successfully developed rust-resistant wheat varieties by incorporating multiple resistance genes (both vertical and horizontal), which have helped to manage disease outbreaks in major wheat-producing regions.

Rice Blast Resistance

Rice blast, caused by *Magnaporthe oryzae*, is one of the most destructive diseases in rice. Breeding for resistance has focused on both race-specific genes like Pi-ta and general resistance genes. Incorporating multiple resistance genes into modern rice varieties has reduced the disease's impact in many rice-growing areas.

Future of Host Resistance in Disease Management

The future of host resistance lies in leveraging advances in genetic technology to develop durable, broad-spectrum resistance. Key trends include:

1. **Pyramiding Resistance Genes:** Combining multiple resistance genes in a single variety to provide durable, broad-spectrum protection.
2. **Genomic Selection:** Using genome-wide information to accelerate the breeding of resistant varieties.
3. **CRISPR-Cas9 Gene Editing:** Precision gene editing to introduce or enhance resistance traits in plants.
4. **Resurrecting Wild Relatives:** Tapping into the rich genetic diversity of wild plant relatives to introduce novel resistance traits.

16

Cultural Methods in Plant Disease Management

Abstract

Cultural methods in plant disease management encompass agronomic practices aimed at creating unfavourable conditions for pathogen development while enhancing crop health and vigour. These sustainable, environmentally friendly techniques form the cornerstone of integrated disease management (IDM). This chapter covers key cultural practices including crop rotation, sanitation, planting time and spacing, irrigation and drainage management, use of resistant varieties, and soil fertility management. By reducing pathogen inoculum, limiting spread, and promoting plant resilience, cultural methods reduce dependency on chemicals and support long-term disease control. The chapter emphasizes the integration of these methods with other control strategies to optimize sustainable disease management.

Keywords: *Cultural methods, plant disease management, crop rotation, sanitation, resistant varieties,*

Cultural methods of plant disease management refer to the application of agronomic practices that help reduce the incidence and severity of diseases. These methods emphasize creating conditions that are unfavourable for pathogen development and spread while promoting the health and vigour of the crop. They are sustainable and environmentally friendly, as they often involve practices that can be integrated into routine crop management.

In this chapter, we will explore various cultural practices, including crop rotation, sanitation, planting time and spacing, irrigation management, and more. These techniques form the foundation of integrated disease management (IDM) and can be used in combination with other control measures for effective disease control.

Importance of Cultural Methods in Disease Management

Cultural practices serve as preventive measures to minimize disease outbreaks by influencing the interaction between the host, pathogen, and environment. By modifying farming practices, it is possible to:

a) **Reduce Pathogen Inoculum:** Prevent the initial development of pathogens in the field.

b) **Limit Pathogen Spread:** Break the life cycle of pathogens and reduce the spread to healthy plants.

c) **Enhance Plant Health:** Promote the overall health of the crop, making it more resistant to diseases.

d) **Reduce Dependence on Chemicals:** Provide a sustainable alternative to chemical control, which helps in reducing costs and environmental risks.

Key Cultural Methods in Plant Disease Management

Several cultural practices have been widely adopted for disease management. These methods are simple, cost-effective, and can be integrated into any farming system.

Crop Rotation

Crop rotation involves planting different crops in a sequence on the same piece of land over different growing seasons. This practice helps break the life cycle of pathogens that are specific to certain crops by depriving them of their preferred host.

How Crop Rotation Helps

a) **Reduces Soil-Borne Pathogens:** Many pathogens, such as fungi and nematodes, persist in the soil and infect successive crops of the same type. By rotating crops with non-host species, the pathogen population decreases.

b) **Reduces Disease Pressure:** Diseases like clubroot in crucifers and root rot in legumes are significantly reduced with crop rotation.

c) **Improves Soil Health:** Diverse crops improve soil structure and fertility, leading to stronger and healthier plants, which are less susceptible to diseases.

Example: In the case of ***Fusarium*** **wilt** in tomatoes, rotating with non-susceptible crops like cereals or corn can reduce the pathogen's soil population, minimizing the risk of infection in the next tomato crop.

Sanitation

Sanitation refers to the removal or destruction of plant residues, infected plant parts, or other materials that may harbour pathogens. It is essential in reducing the amount of pathogen inoculum available for infecting healthy plants.

Key Sanitation Practices

a) **Field Sanitation:** Remove and destroy crop residues, fallen fruits, and infected plant debris after harvesting. These materials often contain pathogens that can survive in the soil or air until the next growing season.

b) **Tool Sanitation:** Regularly disinfect farm tools, machinery, and containers to prevent the spread of pathogens from one field to another.

c) **Greenhouse Sanitation:** Maintain clean conditions in greenhouses by removing dead plant material, disinfecting benches, and ensuring good ventilation.

Example: Sanitation is particularly effective in managing late blight of potatoes and tomatoes, caused by ***Phytophthora infestans***. Removing infected plant debris from the field and practicing crop rotation significantly reduces the incidence of the disease.

Planting Time and Spacing

Adjusting planting time and spacing can be crucial in avoiding disease outbreaks. The timing of planting can be used to escape periods of high disease pressure, while proper spacing ensures good air circulation, reducing the spread of pathogens.

Benefits of Adjusting Planting Time

a) **Avoiding Disease-Conducive Conditions:** Planting early or late, depending on the crop and climate, can help avoid the peak time for pathogen activity. For example, sowing wheat early can help avoid rust infections that are more prevalent in warmer, wetter conditions later in the season.

b) **Reducing Disease Incidence:** Some pathogens, like viruses spread by insect vectors, can be avoided by adjusting planting times to periods when the vectors are less active.

c) **Importance of Proper Spacing:** Improved Air Circulation: Proper spacing between plants allows better airflow, reducing humidity levels and the development of fungal diseases such as powdery mildew and downy mildew.

d) **Reduced Plant-to-Plant Contact:** Dense planting can increase the spread of diseases through contact between infected and healthy plants. Wider spacing minimizes this risk.

Example: In the case of rice blast (caused by *Magnaporthe oryzae*), adjusting planting time and using wider plant spacing can significantly reduce disease incidence, as blast thrives in high humidity and densely planted fields.

Proper Irrigation and Drainage Management

Water management is crucial in preventing many plant diseases, especially those caused by soil-borne and foliar pathogens. Both over-irrigation and poor drainage can lead to the development of diseases.

Key Irrigation Practices

a) **Avoid Overwatering:** Excess water creates conditions favourable for root rot diseases, such as those caused by *Pythium* and *Phytophthora*. Use precise irrigation techniques like drip irrigation to minimize the amount of water applied directly to the soil.

b) **Use of Mulching:** Mulching can help retain soil moisture while reducing splash dispersal of soil-borne pathogens, especially in crops like tomatoes and strawberries.

c) **Timing of Irrigation:** Irrigating in the morning allows plants to dry quickly, reducing the chance of foliar diseases like downy mildew and leaf spot.

Importance of Drainage

a) **Preventing Waterlogging:** Poor drainage can lead to waterlogging, which promotes the growth of root pathogens like *Rhizoctonia* and *Fusarium*. Ensure proper drainage systems are in place to remove excess water.

b) **Reducing Humidity:** Proper drainage prevents the buildup of humidity in the root zone, reducing the risk of diseases caused by fungal pathogens.

   Example: In the management of damping-off disease in seedlings, which is caused by various fungi like *Pythium* and *Rhizoctonia*, proper irrigation techniques, and well-drained soils are essential to prevent pathogen establishment.

Use of Resistant Varieties

Choosing crop varieties that are resistant or tolerant to specific diseases is one of the most effective cultural practices. Resistant varieties have genetic traits that allow them to withstand or suppress pathogen infection.

Benefits of Resistant Varieties

a) **Inherent Disease Control:** Resistant varieties provide protection without additional inputs, reducing the need for fungicides or other treatments.

b) **Sustainability:** The use of resistant varieties promotes long-term disease management by reducing the pathogen population in the field.

Example: The use of downy mildew-resistant cucumber varieties has significantly reduced the need for fungicide applications in many regions, contributing to both economic and environmental benefits.

Soil Management and Fertility

Maintaining healthy soil is essential for preventing many soil-borne diseases. Proper soil fertility and structure promote strong root development and plant health, making plants more resilient to infections.

Key Soil Management Practices:

a) **Soil Amendments:** The addition of organic matter such as compost improves soil structure, drainage, and microbial activity, which helps suppress soil-borne pathogens.

b) **Balanced Fertilization:** Both nutrient deficiencies and excesses can predispose plants to diseases. For example, excessive nitrogen fertilization can promote lush, tender growth, which is more susceptible to diseases like powdery mildew.

c) **pH Management:** Maintaining the appropriate soil pH can suppress certain diseases. For example, adjusting soil pH can help manage clubroot disease in cruciferous crops, which is favored by acidic soils.

Example: Adding organic matter and adjusting the pH to alkaline conditions can reduce the incidence of clubroot in brassicas by creating an environment less conducive to the pathogen.

Integration of Cultural Methods in Disease Management

Cultural methods work best when integrated with other disease management strategies such as biological control, chemical applications, and the use of resistant varieties. This approach, known as Integrated Disease Management (IDM), combines multiple practices to achieve effective and sustainable disease control. Cultural methods form the foundation of IDM by creating unfavourable conditions for pathogens and promoting plant health.

17

Biological Methods of Disease Management

Abstract

Biological methods of disease management utilize living organisms, known as biological control agents (BCAs), to suppress plant pathogens and reduce reliance on chemical pesticides. This chapter explores various BCAs, including antagonistic fungi, bacteria, predatory organisms, and parasitic nematodes, and their modes of action such as predation, parasitism, antibiosis, competition, and induced systemic resistance (ISR). These eco-friendly techniques contribute to sustainable agriculture and form a critical part of integrated disease management (IDM). The chapter highlights successful case studies of biological control in managing diseases like Fusarium wilt, fire blight, and powdery mildew. It also discusses current challenges and future prospects, including advances in biotechnology aimed at improving biocontrol efficacy.

Keywords: *Biological control, plant disease management, biological control agents, induced systemic resistance, microbial biocontrol, disease suppression.*

Biological control in plant pathology refers to the use of living organisms to suppress or control plant pathogens, pests, and diseases, thereby reducing the need for chemical pesticides. These living organisms, known as biological control agents (BCAs), can be naturally occurring or introduced, and they act by various mechanisms to inhibit or eliminate the pathogen.

These methods harness natural enemies of pathogens such as antagonistic fungi, bacteria, or other beneficial organisms to reduce the population of harmful pathogens in the environment. Biological control is an important component of sustainable agriculture, offering an eco-friendly alternative to chemical pesticides and contributing to integrated disease management (IDM).

Key Mechanisms of Biological Control

1. **Predation**: Certain organisms, like *Trichoderma* species or predatory mites, directly attack and consume plant pathogens, reducing their population.

2. **Parasitism**: Some fungi, like *Ampelomyces quisqualis*, parasitize plant pathogens, such as powdery mildew, thereby suppressing their growth and reproduction.
3. **Antibiosis**: Microorganisms like *Bacillus subtilis* and *Pseudomonas fluorescens* produce antibiotics that inhibit the growth or kill plant pathogens. For example, *Bacillus subtilis* produces iturin, which suppresses fungal growth.
4. **Competition**: Certain beneficial microorganisms compete with pathogens for nutrients and space. For instance, beneficial bacteria or fungi occupy ecological niches in the rhizosphere, outcompeting harmful pathogens for vital resources.
5. **Induced Resistance**: Some biological control agents, like *Rhizobacteria*, trigger a plant's natural defense mechanisms, known as induced systemic resistance (ISR), making the plant more resistant to future pathogen attacks.

Benefits of Biological Control

1. **Environmentally Friendly:** Biocontrol methods do not pose the same environmental and health risks as chemical pesticides.
2. **Sustainable:** Using biological control agents can help maintain ecological balance by enhancing biodiversity and reducing chemical inputs.
3. **Pathogen-Specific:** Many biocontrol agents target specific pathogens, minimizing the risk of harming non-target organisms, including beneficial insects and microorganisms.
4. **Integration with Other Methods:** Biological control is compatible with other disease management practices, such as cultural methods and chemical control, making it a key part of IDM strategies.

Types of Biological Control Agents

There are several types of biological control agents used to manage plant diseases, including antagonistic fungi and bacteria, predatory organisms, and parasitic agents.

Antagonistic Fungi

Certain fungi are antagonistic to plant pathogens, meaning they inhibit the growth of or destroy pathogenic fungi. These beneficial fungi work through various mechanisms, including competition, parasitism, and production of toxic compounds.

Example of Antagonistic Fungi: *Trichoderma* species: *Trichoderma* is one of the most widely used biocontrol agents in agriculture. It controls a wide range of soil-borne pathogens, including *Fusarium*, *Pythium*, and *Rhizoctonia*. *Trichoderma* works by:

a) **Competition for Nutrients:** It competes with pathogens for space and nutrients in the soil, effectively starving the harmful fungi.

b) **Mycoparasitism:** Trichoderma directly parasitizes other fungi by penetrating their cell walls and degrading them using enzymes.

c) **Induced Resistance:** It stimulates the plant's natural defenses, making it more resistant to diseases.

Example: In controlling *Fusarium* wilt in crops like tomato and chickpea, *Trichoderma* species have shown excellent results by reducing disease severity and improving crop yield.

Antagonistic Bacteria

Certain bacteria have antagonistic properties against plant pathogens. These bacteria work in the rhizosphere (root zone) or on plant surfaces and suppress pathogens through various mechanisms.

Example of Antagonistic Bacteria: *Bacillus subtilis*: This bacterium is a well-known biocontrol agent that protects plants from pathogens such as *Rhizoctonia*, *Pythium*, and *Alternaria*. It acts by:

a) **Producing Antibiotics:** *Bacillus* produces antimicrobial compounds that inhibit or kill pathogenic fungi and bacteria.

b) **Inducing Systemic Resistance:** It triggers the plant's immune system, leading to systemic resistance against a wide range of pathogens.

c) **Competition:** *Bacillus* competes with pathogens for space and nutrients, limiting the pathogen's ability to establish itself.

Example: *Bacillus subtilis* is used to manage early blight in tomatoes and powdery mildew in cucurbits, effectively reducing disease incidence without harming the environment.

Predatory and Parasitic Organisms

Some beneficial organisms, such as predatory fungi, nematodes, and insects, act as biological control agents by directly attacking plant pathogens or their vectors.

Examples: Nematophagous Fungi: These fungi prey on plant-parasitic nematodes, which can cause significant damage to crops. *Arthrobotrys* and *Paecilomyces* are examples of fungi that trap and kill nematodes, reducing nematode populations and the diseases they transmit.

Parasitic Nematodes

Some nematodes, such as *Steinernema* and *Heterorhabditis*, are used to control soil-dwelling insect pests that act as disease vectors. Example: In controlling root-knot nematodes (*Meloidogyne spp.*), which cause severe damage to the roots of various crops, nematophagous fungi like *Trichoderma* and parasitic nematodes have proven effective in reducing nematode populations and improving crop health.

Mechanisms of Biological Control

Biological control agents manage plant diseases through several mechanisms. Understanding these mechanisms helps in selecting the appropriate agent for a specific pathogen or disease.

Competition for Resources

Biocontrol agents compete with pathogens for essential resources such as nutrients, water, and space. By outcompeting the pathogen, the biocontrol agent prevents it from establishing and reproducing. Example: *Pseudomonas fluorescens*, a beneficial bacterium, competes with pathogenic bacteria and fungi in the rhizosphere for nutrients like iron. By limiting the pathogen's access to nutrients, it reduces the chances of disease development.

Parasitism

Parasitism occurs when a biocontrol agent directly attacks and feeds on the pathogen, leading to its death. Fungal parasites, in particular, can invade the cells of pathogenic fungi, causing them to die. Example: *Trichoderma* species parasitize fungal pathogens by attaching to their hyphae, producing enzymes that degrade the pathogen's cell walls, and eventually killing it.

Antibiosis

Some biocontrol agents produce chemical compounds, such as antibiotics or toxins, that inhibit or kill pathogens. These compounds can suppress the growth or reproduction of the pathogen, preventing it from causing disease. Example: *Bacillus subtilis* produces a range of antibiotics, including bacillomycin and iturin, which are toxic to fungal pathogens like *Botrytis cinerea* (the cause of grey mold in crops like grapes and strawberries).

Induced Systemic Resistance (ISR)

Some biocontrol agents trigger the plant's natural defense mechanisms, making it more resistant to future attacks by pathogens. This process, known as induced systemic resistance (ISR), primes the plant to respond more effectively to pathogen invasion. Example: When plants are treated with *Pseudomonas fluorescens*, it activates ISR, which enhances the plant's ability to fight off diseases such as bacterial wilt in tomatoes and blast in rice.

Success Stories in Biological Control

Several biological control programs have demonstrated remarkable success in managing plant diseases across the world. These examples highlight the potential of biocontrol agents to provide effective and sustainable disease management solutions.

Control of Fire Blight in Apples and Pears

Fire blight, caused by the bacterium *Erwinia amylovora*, is a devastating disease of apples and pears. Biological control using *Pseudomonas fluorescens* and *Bacillus subtilis* has been successfully implemented to reduce fire blight incidence by inhibiting the growth of the pathogen on plant surfaces.

Management of Fusarium Wilt in Tomatoes

Fusarium wilt, caused by *Fusarium oxysporum*, is a serious problem for tomato growers. The use of *Trichoderma* species as a soil treatment has significantly reduced disease severity and improved yields. *Trichoderma* parasitizes the *Fusarium* fungus and enhances the plant's root health, making it more resilient to infection.

Biological Control of Powdery Mildew in Cucurbits

Powdery mildew, a common fungal disease in cucurbits, has been effectively managed using *Ampelomyces quisqualis*, a mycoparasite that infects and destroys the mildew spores. This biocontrol agent has been successfully applied in organic and conventional farming systems to reduce the need for chemical fungicides.

Challenges and Future Prospects

While biological control offers many benefits, it also faces challenges, including inconsistent performance under different environmental conditions and limited shelf life of biocontrol products. Additionally, biocontrol agents may take longer to act than chemical treatments, which can be a limitation in severe disease outbreaks.

However, ongoing research is improving the effectiveness and reliability of biocontrol methods. Advances in biotechnology, including the development of genetically modified organisms (GMOs) and microbial consortia, hold promise for enhancing the efficacy of biocontrol agents.

18

Integrated Disease Management (IDM)

Abstract

Integrated Disease Management (IDM) is a holistic and sustainable approach to managing plant diseases by combining biological, chemical, cultural, and physical control methods in a coordinated manner. IDM aims to keep disease levels below economic thresholds while minimizing the reliance on chemical pesticides, thus preserving environmental health and promoting sustainable agriculture. Key principles include prevention, use of diverse control tactics, regular monitoring, minimizing chemical inputs, and maintaining ecological balance. The chapter discusses the components of IDM, including the role of biological agents such as Trichoderma, Bacillus subtilis, and Pseudomonas fluorescens; judicious use of chemicals; cultural practices like crop rotation and sanitation; and physical methods such as soil solarization and seed treatments. The disease triangle concept is applied to illustrate IDM strategies targeting the host, pathogen, and environment. Case studies on rice blast and potato late blight demonstrate the practical application and success of IDM programs. Despite challenges like farmer education, cost, and pathogen evolution, advances in technology, genomics, and microbial consortia present promising future directions for IDM.

Keywords: *Integrated Disease Management (IDM), Induced systemic resistance (ISR), Pathogen management, Disease monitoring, Environmental sustainability*

Integrated Disease Management (IDM) is an approach to controlling plant diseases that combines multiple management strategies to minimize the risk of disease development and spread. IDM integrates biological, chemical, cultural, and physical control methods in a coordinated and complementary way. The primary goal of IDM is to keep disease levels below economically damaging thresholds while reducing reliance on chemical inputs, preserving environmental quality, and promoting sustainability in agricultural systems.

Concept of Integrated Disease Management (IDM)

IDM is designed to harmonize various disease control measures in a way that limits disease progression without relying heavily on any single method. Its goal is to maintain disease levels below an economic threshold where the cost of intervention is justified by the gains in crop yield and quality. IDM emphasizes preventing disease outbreaks through the judicious use of all available management tools, ensuring that environmental and human health risks are minimized.

Key Principles of IDM

1. **Prevention over Cure:** The cornerstone of IDM is prevention. By managing conditions that favor pathogen development, it is possible to reduce the inoculum load and avoid disease outbreaks. Prevention strategies include crop rotation, resistant varieties, proper sanitation, and maintaining soil health.
2. **Diverse Control Tactics:** A variety of control methods biological, chemical, cultural, and physical are used in combination. Each method targets different stages of the pathogen lifecycle to ensure comprehensive management. This also reduces the risk of pathogen resistance to any one tactic.
3. **Regular Monitoring and Informed Decision-Making:** Continuous crop monitoring helps in identifying early signs of disease. Decisions regarding intervention are based on field observations, weather forecasts, disease severity, and economic thresholds, ensuring timely and effective actions.
4. **Minimizing Chemical Use:** Although chemical control is an important part of IDM, its use is minimized and carefully managed. Chemicals are typically used as a last resort or in combination with other methods, reducing their environmental impact and delaying pathogen resistance.
5. **Ecological Balance:** IDM maintains a balance between effective disease management and protecting non-target organisms, including beneficial insects, soil microorganisms, and natural predators of pests.

Components of IDM

The key components of IDM biological, chemical, cultural, and physical work synergistically to control plant diseases. Each component has unique strengths, and their combined use offers a more robust solution than any individual tactic.

Biological Methods

Biological control in plant pathology refers to the use of living organisms, such as natural predators, parasites, pathogens, or competitors, to control plant diseases and reduce the impact of harmful pathogens. This method leverages the natural interactions between organisms to suppress the populations of disease-causing agents in plants. Biological control is an eco-friendly, sustainable alternative to chemical pesticides and is an important component of Integrated Disease Management (IDM).

***Trichoderma* spp.:** This beneficial fungus is widely used in soil to suppress soil-borne pathogens like *Fusarium*, *Rhizoctonia*, and *Pythium* through mechanisms such as competition, mycoparasitism, and enzyme production.

***Bacillus subtilis*:** A bacterium used to control various foliar and soil-borne diseases, including *Rhizoctonia* and *Sclerotinia*, by producing antibiotics and inducing systemic resistance in plants.

***Pseudomonas fluorescens*:** This bacterium induces systemic resistance in crops and is particularly effective in controlling root pathogens such as *Phytophthora* and *Pythium*.

Chemical Methods

Chemical control involves the judicious use of fungicides, bactericides, and nematicides to control pathogens. While necessary in many cases, chemical control is used with caution in IDM to minimize environmental and human health risks. The goal is to integrate chemicals into a broader disease management plan that also includes biological, cultural, and physical methods.

Guidelines for chemical use in IDM

a) **Timing and Targeted Application:** Chemicals are applied based on disease forecasts and early warning systems, ensuring they are used only when necessary. This reduces overuse and prevents the buildup of resistant pathogen strains.

b) **Rotational Use:** Different classes of chemicals are rotated to prevent pathogen resistance. For example, alternation between fungicides with different modes of action (e.g., triazoles and strobilurins) is common in managing fungal diseases like powdery mildew or downy mildew.

c) **Compatibility with Biocontrol:** Chemicals are selected carefully to ensure they do not harm beneficial organisms used in biological control.

Cultural Methods

Cultural practices are fundamental to IDM, creating an environment that is unfavorable to pathogens while promoting healthy plant growth. These methods focus on modifying the cropping environment and crop management practices to reduce disease risk.

a) **Crop Rotation:** Rotating different types of crops can break the life cycles of soil-borne pathogens. For example, rotating cereals with legumes reduces the buildup of pathogens like *Verticillium* and *Fusarium*.

b) **Sanitation:** Removing plant debris, disinfecting tools, and using pathogen-free planting material helps to eliminate sources of inoculum. This practice is especially important in greenhouses and high-value crops like vegetables and fruits.

c) **Optimal Planting Time:** By adjusting planting dates, farmers can avoid periods when environmental conditions favor pathogen activity. For example, sowing crops before or after periods of high humidity can help manage diseases like rusts and downy mildew.

d) **Irrigation Management:** Proper water management is essential to avoid conditions like waterlogging, which can promote diseases such as *Phytophthora* and *Pythium* root rots.

Physical Methods

Physical methods rely on non-chemical and non-biological interventions to manage pathogens. These methods act by either directly reducing pathogen populations or altering the environment to be less conducive to pathogen survival.

a) **Soil Solarization:** Clear plastic is used to cover moist soil during periods of intense sunlight, trapping heat that kills soil-borne pathogens. Solarization is especially effective against fungi like *Verticillium*, *Fusarium*, and nematodes.

b) **Mulching and Netting:** Mulches can prevent soil from splashing onto plants, thereby reducing the spread of soil-borne pathogens like *Fusarium*. Netting can act as a barrier to insect vectors that transmit viral diseases.

c) **Hot Water Treatment:** Seeds, tubers, or bulbs can be treated with hot water to kill pathogens before planting. This method is often used for crops like potatoes and bulbs to control fungal and bacterial diseases.

The Disease Triangle in IDM

The disease triangle is a conceptual model that represents the three factors necessary for disease development: the host, pathogen, and environment. Disease can only occur when all three factors are favorable for pathogen infection and proliferation. IDM strategies target one or more sides of this triangle to reduce disease outbreaks.

a) **Host Resistance:** Selecting or breeding resistant plant varieties is one of the most effective strategies in IDM. Resistant varieties limit the ability of pathogens to infect the plant.

b) **Pathogen Control:** Pathogen populations are reduced through various means such as biological control, chemical control, or sanitation practices.

c) **Environmental Management:** By modifying environmental conditions (e.g., irrigation, planting date), it is possible to reduce the favorable conditions that support pathogen development.

Case Studies of Successful IDM Programs

a) IDM for Rice Blast Disease

1. Rice blast caused by *Magnaporthe oryzae* is a major threat to rice cultivation. IDM strategies for managing this disease include:
2. Resistant Varieties: Breeding and planting rice varieties with genetic resistance to blast.
3. Biological Control: Application of *Pseudomonas fluorescens* to induce systemic resistance in rice.
4. Cultural Practices: Altering planting times to avoid high-humidity periods that favor infection.
5. Chemical Control: Fungicides like triazoles applied at early infection stages to limit disease spread.

b) IDM for Late Blight in Potatoes

1. Late blight, caused by *Phytophthora infestans*, can cause severe yield losses in potatoes. A comprehensive IDM approach includes:
2. Resistant Varieties: Developing and planting varieties with resistance to late blight.
3. Cultural Practices: Regular removal of infected plant debris and crop rotation to interrupt the pathogen's lifecycle.
4. Fungicide Rotation: Using fungicides with different modes of action in rotation to prevent resistance development.
5. Biological Control: *Trichoderma harzianum* is applied to soil to manage the soil-borne stages of the pathogen.

Challenges and Future Prospects of IDM

While IDM offers a sustainable approach to disease management, it faces challenges, including access to resources, knowledge dissemination, and pathogen adaptation.

Challenges

1. Farmer Knowledge and Training: Effective IDM requires access to education and extension services so that farmers can understand and implement diverse strategies.
2. Costs: Some IDM methods, particularly biological control products, may be costly or difficult to access in certain regions.
3. Pathogen Evolution: Pathogens may adapt, becoming resistant to control measures or biocontrol agents, requiring constant adaptation of IDM strategies.

Future Prospects

1. Technological Advancements: Improved disease detection, forecasting models, and precision agriculture techniques will make IDM more efficient and accessible.
2. Genomics and Breeding: Understanding the genetic makeup of both pathogens and crops will lead to the development of new resistant varieties.
3. Microbial Consortia: Advances in biological control could see the development of microbial consortia combinations of different beneficial organisms that provide more stable and effective disease control.

References

Agrios, G.N. 2010. Plant Pathology. Acad. Press.
Alexopoulos, Mims and Blackwel. Introductory Mycology.
Dhingra, O.D. and Sinclair, J.B. 1986. Basic Plant Pathology Methods. CRC Press, London, Tokyo.
Gibbs, A. and Harrison, B. 1976. Plant Virology - The Principles. Edward Arnold, London
Goto, M. 1990. Fundamentals of Plant Bacteriology. Academic Press, New York.
Hull R. 2002. Mathew's Plant Virology. 4th edn. Academic Press, New York.
Kamat, M. N. Introductory Plant Pathology. Prakash Pub, Jaipur.
Mehrotra, R.S. and Aggarwal, A. 2007. Plant Pathology. 7th edn. Tata Mc Graw Hill Publ. Co. Ltd.
Nene, Y.L. and Thapliyal, P.N. 1993. Fungicides in Plant Disease Control. 3rd Ed. Oxford & IBH, New Delhi.
Pathak, V. N. Essentials of Plant Pathology. Prakash Pub., Jaipur
Rajeev, K. and Mukherjee, R.C. 1996. Role of Plant Quarantine in IPM. Aditya Books.
Rhower, G.G. 1991. Regulatory Plant Pest Management. In: Handbook of Pest Management in Agriculture. 2nd edn. Vol. II. (Ed. David Pimental). CRC Press.
Singh R.S. 2008. Plant Diseases. 8 th Ed. Oxford & IBH. Pub. Co.
Singh R.S. 2013. Introduction to Principles of Plant Pathology. Oxford and IBH Pub. Co.
Verma, J.P. 1998. The Bacteria. Malhotra Publ. House, New Delhi.
Vyas SC. 1993. Handbook of Systemic Fungicides. Vols. I-III. Tata McGraw Hill, New Delhi.

Glossary

Adjuvants: Additives mixed with fungicides to enhance their performance. These are usually inactive substances, such as surfactants, that improve how the chemical spreads, sticks, or penetrates plant surfaces by altering surface tension.

Antagonism: A relationship where one microorganism harms or inhibits the growth of another, often through competition or by producing harmful substances.

Antagonists: Organisms that can suppress or interfere with the growth of other microorganisms, often used in biological control to reduce plant diseases.

Antibiosis: A form of microbial warfare where one organism produces a chemical that negatively affects another organism's growth or survival.

Appressorium: A swollen structure at the tip of a fungal germ tube or hypha that helps the fungus stick to and break into the plant's surface.

Avirulent (Non-virulent): Describes a pathogen that cannot cause disease in a particular host plant.

Bactericide: A chemical substance used specifically to kill or inhibit bacteria.

Bacteriology: The scientific study of bacteria, including their role in plant health and disease.

Biotroph: A typc of organism that survives and reproduces only by feeding on living plant tissue.

Compatibility: The ability of two or more substances (like pesticides or chemicals) to be mixed or used together without reducing their effectiveness or causing negative reactions.

Competition: When two organisms try to use the same resource like nutrients or space, the one that uses it more effectively dominates.

Culture: The act of growing microorganisms on a prepared nutrient medium under controlled conditions for research or diagnosis.

Disease Cycle: The complete series of steps a disease goes through from survival of the pathogen to infection, colonization, reproduction, and spread.

Disease Incidence: The number or proportion of plants in a population showing disease symptoms.

Disease Severity: A measure of how much damage a disease has caused, often expressed as a percentage of affected plant tissue or yield.

Disinfectant: A chemical or physical agent that eliminates pathogens from plant tissues, tools, or surfaces.

Disorder: A plant problem not caused by a pathogen, but by non-living (abiotic) factors such as nutrient imbalance, pollution, or weather extremes.

Eradication: Controlling disease by removing or destroying the pathogen or infected plant parts to prevent further spread.

Exclusion: Preventing the introduction of pathogens into a clean area, often through quarantine or sanitation practices.

Fungicide: A chemical compound that kills or inhibits the growth of fungi.

Hyper-parasitism: When one parasite infects or lives off another parasite. For example, some fungi can parasitize other fungi.

Hyperplasia: Abnormal enlargement of plant tissues or organs caused by an increase in the number of cells.

Hypersensitivity: A rapid and intense reaction of plant cells to infection, where affected cells die quickly to block the pathogen's spread.

Hypertrophy: Abnormal growth of plant tissues due to an increase in cell size, rather than number.

Hypha (plural: Hyphae): A single, thread-like filament of a fungus. Together, many hyphae form the fungal body.

Immune: Completely resistant to a specific pathogen; the organism cannot be infected.

Infection: The successful entry and establishment of a pathogen inside a plant.

Infectious Disease: A disease caused by a pathogen that can spread from one plant to another.

Inoculate: To deliberately introduce a pathogen to a plant to study its effect or for diagnostic purposes.

Inoculation: The process by which a pathogen comes into contact with a potential host plant.

Inoculum: The actual part of the pathogen (like spores or bacteria) that can start an infection.

Inoculum Potential: A combination of the amount of inoculum present and its ability to cause infection.

Isolate: A pure sample of a microorganism obtained from a single spore or infected plant.

Isolation: The process of separating a pathogen from its host and growing it on a nutrient medium for study.

Latent Infection: An infection where the pathogen is present in the plant, but there are no visible symptoms.

Microscopic: So small that it can only be seen with the help of a microscope.

Mycelium (plural: Mycelia): A mass of fungal hyphae that collectively make up the body of a fungus.

Mycology: The branch of biology that deals with the study of fungi.

Necrosis: Death of plant tissues, often seen as brown or black patches on leaves or stems.

Necrotroph: A type of pathogen that kills plant cells and then feeds on the dead tissue.

Nematicides: Chemicals used to kill or control nematodes (microscopic, worm-like pests).

Nematology: The scientific study of nematodes, especially those affecting plants.

Non-Infectious Disease: Plant disorders caused by non-living factors like nutrient deficiency, chemicals, or temperature, not by pathogens.

Pathogenesis: The series of events that occur as a pathogen infects and causes disease in a plant.

Pathogenicity: A pathogen's ability to cause disease in a particular host.

Penetration: The initial entry of a pathogen into the plant's tissues.

Phyllody: A condition in which floral parts transform into green, leafy structures due to pathogen interference.

Phytotoxicity: A harmful effect caused to plants by chemicals, such as pesticides, often resulting in burn or stunted growth.

Plant Disease: Any abnormal condition in a plant that harms its growth, structure, or economic value.

Primary Infection: The first infection in a growing season, usually caused by overwintering or over summering pathogens.

Primary Inoculum: The part of the pathogen that survives between growing seasons and causes the first infections.

Quarantine: Regulations that restrict the movement of plants or plant products to prevent the spread of diseases.

Resistant: A plant's ability to slow down or stop the development of a pathogen after infection.

Saprophyte: An organism that lives on and feeds from dead or decaying organic matter.

Secondary Infection: Infection caused by the spread of pathogens during the same season, following a primary infection.

Secondary Inoculum: The spores or other infectious units produced from earlier infections in the same crop cycle.

Sign: The actual presence of the pathogen or its parts (like fungal spores or bacterial ooze) visible on the plant.

Spore: A tiny reproductive unit produced by fungi and some other organisms, often involved in spreading disease.

Susceptible: A plant that lacks defense mechanisms and is easily infected by a specific pathogen.

Symptom: Visible or measurable effects of disease on a plant, like spots, wilting, or discoloration.

Syndrome: A group of symptoms that consistently occur together and characterize a specific plant disease.

Tenacity: The ability of a pesticide or chemical to stick to plant surfaces and resist being washed off.

Tolerance: The plant's capacity to endure a disease with minimal impact on yield or health, even if infection occurs.

Transmission: The spread of a pathogen from one plant to another, often via wind, water, insects, or tools.

Vector: Any living organism (e.g., insects, mites, nematodes) that carries and transmits a pathogen between plants.

Virology: The scientific study of viruses and their interactions with host organisms.

Virulence: The degree to which a pathogen can cause disease; a measure of its aggressiveness.

Virulent: Describes a pathogen that is highly capable of causing serious disease.

Virus: A sub-microscopic infectious agent made of genetic material (DNA or RNA) enclosed in a protein coat. It cannot reproduce without a host.

Index